PEIRCE'S THEORY OF ABDUCTION

PEIRCE'S THEORY OF ABDUCTION

by

K. T. FANN

MARTINUS NIJHOFF / THE HAGUE / 1970

ISBN-13: 978-90-247-0043-1 e-ISBN-13: 978-94-010-3163-9
DOI: 10.1007/978-94-010-3163-9

PREFACE

This monograph attempts to clarify one significant but much neglected aspect of Peirce's contribution to the philosophy of science. It was written in 1963 as my M.A. thesis at the University of Illinois. Since the topic is still neglected it is hoped that its publication will be of use to Peirce scholars.

I should like to acknowledge my indebtedness to Dr. Max Fisch who broached this topic to me and who advised me continuously through its development, assisting generously with his own insights and unpublished Peirce manuscripts.

TABLE OF CONTENTS

INTRODUCTION

1. A CURRENT ISSUE IN THE PHILOSOPHY OF SCIENCE

Is there a logic of scientific discovery? If so, what is the nature of such a logic? Is the process of conceiving a new idea a reasonable affair subject to a logical investigation or is it purely an irrational "hunch" of an investigator? This is one of the most important current issues in the philosophy of science.

The majority of philosophers deny that there is any logic in proposing a hypothesis. For them the logic of discovery (if it can be properly called such) can only be concerned with the investigation of the methods of testing hypotheses which have already been presented to us. Thus Popper argues: "The initial stage, the act of conceiving or inventing a theory, seems to me neither to call for logical analysis nor to be susceptible of it. The question how it happens that a new idea occurs to a man... may be of great interest to empirical psychology; but it is irrelevant to the logical analysis of scientific knowledge."[1] Braithwaite writes: "There are historical problems, both as to what causes the individual scientist to discover a new idea; and as to what causes the general acceptance of scientific ideas. The solution of these historical problems involves the individual psychology of thinking and the sociology of thought. None of these questions are our business here."[2] Wisdom elaborates: "There is no rational machinery for passing from observational premises to an inductive generalization but that hypothesis is

[1] Karl R. Popper, *The Logic of Scientific Discovery*, (New York: Basic Books, 1959), pp. 20f.

[2] R. B. Braithwaite, *Scientific Explanation*, (Cambridge: Cambridge University Press, 1955), pp. 31f.

attained by some mental jump."[3] And, Copi concludes, "Logic has nothing to say about the *discovery* of hypotheses; this process is more properly to be investigated by psychologists."[4]

These philosophers regard the discovery of new ideas as a mere guess, chance, insight, hunch or some mental jump of the scientist's which is only open to historical, psychological, or sociological investigation. When they talk about the logic of discovery, they are only concerned with the testing of hypotheses. Thus, Wisdom argues: "Observations...instead of leading to a hypothesis, their function is to test it, and the only way of continuing scientific activity is by means of the hypothetico-deductive system."[5] In the hypothetico-deductive system the hypothesis is expressed in a general statement, observations are made and expressed in a particular statement, and a conclusion is deduced and tested by experience.[6] Wisdom goes on to declare that "there is no type of scientific inference that cannot be fitted into the hypothetico-deductive system."[7]

Other philosophers feel that the proposing of a new hypothesis is not a mere guess or hunch, but that there is a logical relation from observations to the new hypothesis. However, they insist that the logic of discovery is nothing more than "a logic of inductive inference"[8] or "a kind of induction."[9] Reichenbach sums up the arguments as follows: "There is an inductive relation from the known facts to the new theory... We shall never have a definite proof of the theory; the so-called confirmation consists in the demonstration of some facts which

[3] John O. Wisdom, *Foundations of Inference in Natural Science*, (London, 1952), p. 49.

[4] Irving M. Copi, *Introduction to Logic*, (New York, Macmillan, 1953), p. 407.

[5] Wisdom, *op. cit.*, p. 51.

[6] Cf. Popper, *op. cit.*, p. 12; Braithwaite, *op. cit.*, Chapter 2; Wisdom, *op. cit.*, p. 51.

[7] Wisdom, *op. cit.*, p. 53.

[8] Cf. F. C. S. Schiller, "Scientific Discovery and Logical Proof," *Studies in the History and the Methods of the Sciences*, (Charles Singer, ed. [Oxford: Clarendon Press, 1917]), Vol. I.; cf. Hans Reichenbach, *Experience and Prediction*, (Chicago: University of Chicago Press, 1938), pp. 381 ff.

[9] Cf. William Kneale, "Induction and Transcendent Hypothesis," in *Readings in the Philosophy of Science*, (H. Feigl, ed. [New York: Appleton-Century-Crofts, 1953]), pp. 353 ff.

confer a higher probability upon the theory, i.e., which allow rather simple inductive inference to the theory."[10] He thinks of induction as the only means for an expansion of knowledge and states: "The methods of induction. . . always will remain the genuine methods of scientific discovery."[11]

A few philosophers have come to regard the process of constructing and selecting a hypothesis as a reasonable affair which is susceptible of a logical analysis. They feel that in scientific discovery, there may be more problems for the logician than just analyzing the arguments supporting already invented hypotheses. Thus, Peirce wrote, "each chief step in science has been a lesson in logic" (5.363).[12] He apparently felt that there is a *conceptual* inquiry, one properly called "a logic of discovery," which is not to be confounded with the psychology, sociology and history of discovery. But most contemporary philosophers are unreceptive to this view, giving most of their attention to inductive reasoning, probability, and the principles of theory construction. Hanson, a staunch supporter of Peirce's view, writes: "But, for Peirce, the work of Popper, Reichenbach, and Braithwaite would read less like a *Logic of Discovery* than like a *Logic of the Finished Research Report.* Contemporary logicians of science have described how one sets out reasons in support of a hypothesis once proposed. They have said nothing about the conceptual context within which such a hypothesis is initially proposed."[13]

One thing should be made clear. When Popper, Braithwaite and Reichenbach[14] urge that there is no logical analysis appropriate to the actual thinking process in scientific discovery,

[10] Reichenbach, *op. cit.*, p. 383.

[11] *Ibid.*

[12] All numerical references refer to the *Collected Papers of Charles Sanders Peirce,* volumes one through six edited by Charles Hartshorne and Paul Weiss (Cambridge, Massachusetts, 1931-1935), volumes seven and eight edited by Arthur Burks, (Cambridge, Massachusetts, 1958). Thus, 5.363 means Volume 5, paragraph 363 of the *Collected Papers.*

[13] N. R. Hanson, "Is There a Logic of Scientific Discovery?" H. Feigl and G. Maxwell, eds., *Current Issues in the Philosophy of Science,* (New York, 1959), p. 21.

[14] Reichenbach, *op. cit.,* p. 5. "Epistemology does not regard the process of thinking in their actual occurrence; this task is entirely left to psychology."

they are saying nothing which Peirce or Hanson would reject. Peirce did not think of himself as writing manuals to help scientists make discoveries. There could be no such thing. What Peirce wanted to insist upon was that the birth of new ideas could never be satisfactorily cleared up by psychological, sociological and historical investigations alone. One important task of a philosopher is to conduct a logical (conceptual) investigation of discovery. There can be good reasons, or bad, for suggesting one kind of hypothesis rather than some other kind. These reasons may be entirely different in type from those which lead one to accept a hypothesis. Peirce wished to show that reasoning *towards* a hypothesis is of a different kind than reasoning *from* a hypothesis. He realized that the former "has usually been considered either as not reasoning at all, or as a species of Induction."[15] But, he said: "I don't think the adoption of a hypothesis on probation can properly be called induction; and yet it is reasoning" (8.388).

In the spirit of Peirce's thesis, Hanon distinguishes:[16]

1. Reasons for accepting a hyppothesis H, from
2. Reasons for suggesting H in the first place.

Our reasons for accepting H would be those which we might have for thinking H true whereas the reasons for suggesting H initially may be those which make H a plausible type of hypothesis. Hanson points out that the difference between (1) and (2) is one of logical type. What are and are not good reasons for adopting a hypothesis on probation is a logical matter, which may be decided on conceptual grounds. No further observations or experiments are required to settle such issues. Most philosophers only concern themselves with analyzing reasons for accepting a hypothesis. "They begin with the hypothesis as given, as cooking recipes begin with the trout,"[17] writes Hanson. To study only the verification of hypotheses

[15] C. S. Peirce, *Letters to Lady Welby,* (Irwin Lieb, ed. [New York: Whitlock's, 1953]), p. 42.

[16] N. R. Hanson, *op. cit.,* p. 22. See also his *Patterns of Discovery,* pp. 85-92; "The Logic of Discovery," *Journal of Philosophy,* LV, 25, pp. 1073-1089; "More on the Logic of Discovery," *Journal of Philosophy,* LVII, 6, pp. 182-188.

[17] Hanson, "Is There a Logic . . .," *op. cit.,* p. 31.

leaves a vital question unanswered – namely, how hypotheses are "caught." "Natural scientists do not 'start from' hypotheses. They start from data."[18] Peirce's theory of abduction[19] is concerned with the reasoning which starts from data and moves towards hypotheses.

2. PEIRCE AND HIS THEORY OF ABDUCTION[20]

Peirce struggled over more than fifty years to lay bare the logic by which we get new ideas (7.98). He certainly was more perceptive and bolder than anyone before or since. Abduction is a distinct type of reasoning (inference or argument), which is not to be confused with the two traditionally recognized types, induction and deduction. Peirce considered abduction to be the essence of his pragmatism (5.196). He insisted that it was essential to history (6.606, 2.714), that it constituted the first stage of all inquiries (6.469), and that it was a necessary part of perception (5.181) and memory (2.625).

Readers of Peirce can hardly help but feel the importance of Peirce's theory of abduction in his philosophy, but it is by no means clear what the nature of abduction is. His writings on this subject are typically fragmentary and, as a consequence, we find many different views represented. However, we must not conclude from this fact that his thought was likewise fragmentary and full of contradictions. It is true that a systematic treatment is not to be expected, nor must we look for uniform consistency in his discussions, which extended over more than half a century. But the lack of unity in his thought has been greatly exaggerated because of a failure to recognize the following facts: (1) His logic is foundational to the rest of his philosophy; (2) His theory of abduction is a highly original investigation; and (3) There was a temporal development in his thought.

[18] *Ibid.*, p. 29.

[19] At different times he called it by different names: Hypothesis, Hypothetic Inference, Retroduction or Presumption. He seemed to prefer *Abduction* as the best designation for the sake of neat enumeration.

[20] See Arthur Burks, "Peirce's Theory of Abduction," *Philosophy of Science*, XIII (1946), pp. 301-06, and Harry G. Frankfurt, "Peirce's Notion of Abduction," *Journal of Philosophy*, VXV, (1958), pp. 593-597.

(1) Peirce preferred to call himself a "logician." Certainly, the study of logic was his grand passion which over-shadowed all others. As a system builder, Peirce considered logic to be the foundation of his philosophy. Murphy observes: "Since logic is the basis of the architectonic order, the creative or dynamic agent in the development of Peirce's philosophy should have been his logic."[21] It is a very important task of logic, Peirce held, to classify reasonings so as to determine the validity of each kind. One of his earliest published papers was concerned with the classification of arguments.[22] In 1910 he said, "I have constantly since 1860, or 50 years, had this question (classification of reasonings) prominently in mind..." (7.98). The question remained a prominent one in his mind until the end of his life.

(2) Peirce regarded himself as "An explorer upon untrodden ground" (2.102) concerning abduction. There was no definite path to be followed. He had to try different paths at different times and as a consequence his terminology varied from time to time. Also, Peirce's discussions on abduction are often connected with his discussions on other topics. His interest concerning abduction in connection with "The Law of Mind" is quite different from that in connection with his discussions on Pragmatism, for example.

(3) This last point is particularly important for an understanding of Peirce's varied writings on abduction. His different views should not be treated as inconsistent theories resulting from the lack of unity in his thought; rather, they should be treated as Peirce's repeated attempts at working out a theory. A proper understanding of Peirce's theory of abduction requires a close attention to the chronology of his writings. He himself said in 1902, concerning his pragmatic maxim of 1877, that "my opinion remains substantially the same now as then but all those years have not passed without my learning anything" (2.99). In his writings on abduction he often referred back to

[21] M. G. Murphey, *The Development of Peirce's Philosophy* (Cambridge: Harvard University Press, 1961), p. 3.

[22] "On the Natural Classification of Arguments" was written in 1867 for the *Proceedings of the American Academy of Arts and Sciences,* (2.461).

his earlier views and pointed out where he had "committed an error," how he was "confused," and where his opinion had "wavered."

It is the purpose of this thesis to elucidate Peirce's theory of abduction by tracing the development of his thought, and by pinpointing his own statements concerning his changes of view. We want to know what Peirce thought to be sufficiently distinctive about abduction to warrant calling it a kind of reasoning. Since most philosophers still consider the adoption of hypotheses as not reasoning at all, or as a species of induction, we want to consider Peirce's conception of reasoning and logic in order to decide whether Peirce was justified in calling abduction a kind of reasoning. Should the theory of the economy of research, which is intimately connected with the suggestion of a hypothesis, be included as a part of logic? Is the process of suggesting a hypothesis initially concerned with hypothesis formation or hypothesis selection, or both? We also want to know what the validity of abduction is and how it is related to deduction and induction. These are the main problems the author intends to discuss in this monologue. But, first of all, let us take a brief glance at the general character of abduction in the following section.

3. THE GENERAL CHARACTER OF ABDUCTION

One task of logic is to classify inferences. Peirce's classification is as follows:

- Inference
  - Explicative (analytic or deductive)
  - Ampliative (synthetic)
    - Abductive
    - Inductive

In explicative inference the conclusion follows from the premises necessarily while in ampliative inference the conclusion does not follow from the premises with necessity. The conclusion amplifies rather than explicates what is stated in the premises. All the empirical sciences use such reasoning. Moreover, it is the only kind of reasoning that introduces new ideas into our store of knowledge.

Peirce's classification is different from the traditional classification because it includes a novel type of inference in addition to induction and deduction. Most logicians identify induction with synthetic reasoning. They failed to recognize the trichotomy of inferences because, Peirce thinks, they had too "narrow and formalistic a conception of inference" (8.228). These logicians generally confine their investigation of reasoning to its 'correctness,' "by which they mean its leaving an absolute inability to doubt the truth of a conclusion so long as the premisses are assumed to be true" (8.383). This amounts to confining their study to deduction. Peirce thinks logicians should have two principal aims: "Ist, to bring out the amount and kind of *security* (approach to certainty) of each kind of reasoning, and 2nd, to bring out the possible and esperable uberty, or value in productiveness, of each kind" (8.384). He insists that ampliative reasonings are of two kinds: Induction and Abduction. Abduction is concerned with the reasons for adopting a hypothesis. The adoption of a hypothesis on probation cannot properly be called induction; "and yet it is *reasoning* and though its *security* is low, its *uberty* is high" (8.388). Thus, from deduction to induction and to abduction the security decreases greatly, while the uberty increases greatly.

Broadly speaking, abduction covers "all the operations by which theories and conceptions are engendered" (5.590). These operations are best manifested in the process of arriving at scientific hypothesis. Peirce thinks this process is essentially inferential. "Although it is very little hampered by logical rules, nevertheless it is logical inference, asserting its conclusion only problematically or conjecturally, it is true, but nevertheless having a perfectly definite logical form" (5.188). Its form is:

> The surprising fact C is observed,
> But if A were true, C would be a matter of course;
> Hence, there is reason to suspect that A is true (5.189).

Such a process is inferential because the hypothesis "is adopted for some reason, good or bad, and that reason, in being regarded as such, is regarded as lending the hypothesis some plausibility." (2.511n.).

It is by no means clear what the "operations by which theories and conceptions are engendered" consist in. Is abduction concerned with the process by which we get new ideas, "the process of forming an explanatory hypothesis" (5.171), or is it concerned with the process of deciding whether an idea is worth testing, "the adoption of an hypothesis on probation" (8.388)? Peirce himself did not always keep this distinction in mind. An attempt will be made to clarify this matter.

In order to understand the nature of abduction it is necessary to compare and connect it with the other two kinds of reasoning. However, since Peirce's views on deduction and induction are well known, an attempt to review his detailed expositions of them will not be made here; they will be mentioned only in so far as they are pertinent to the general theme of the discussion.

Most writers on Peirce's theory of abduction divide Peirce's thought roughly into two periods. The transition from one view to another was made around the turn of the century, but since this transition took place over a period of time it is difficult to pinpoint a definite year. Peirce himself wrote in 1910, "in almost everything I printed before the beginning of this century I more or less mixed up hypothesis [abduction] and induction" (8.227). Writers on Peirce vary widely on this point. Judging from Peirce's writings, the best account is given by Burks; he named the year 1891, when Peirce had retired to his home near Milford, Pennsylvania, as the beginning of a transitional decade dividing the two periods.[23]

In his earlier papers Peirce treated inference, and hence abduction, as an evidencing process. The three types of inferences were considered separate and independent forms of reasoning. Induction "infers the existence of phenomena such as we have observed in cases that are *similar*," while abduction "supposes something of a different kind from what we have directly observed, and frequently something which it would be impossible for us to observe directly" (2.640). For in induction we

[23] Burks, *op. cit.*, p. 301.

generalize from a number of cases of which something is true and infer that the same thing is probably true of a whole class. But in abduction we pass from the observation of certain facts to the supposition of a general principle to account for the facts. Thus induction may be said to be an inference from a sample to a whole, or from particulars to a general law; abduction is an inference from a body of data to an explaining hypothesis, or from effect to cause, "The former classifies, the latter explains" (2.636).

In papers written after 1891 Peirce widened the concept of inference to include methodological process as well as evidencing process. The three kinds of reasoning, while remaining distinguishable, became closely interlinked. Abduction furnishes the reasoner with hypothesis while induction is the method of testing and verifying (2.776). Peirce conceived of the three kinds of reasoning as three stages of inquiry. Abduction invents or proposes an hypothesis; it is the initial proposal of an hypothesis on probation to account for the facts. Deduction explicates hypotheses, deducing from them the necessary consequences which may be tested. Induction consists in the process of testing hypotheses. Thus, "Abduction is the process of forming an explanatory hypothesis. It is the only logical operation which introduces any new ideas; for induction does nothing but determine a value, and deduction merely evolves the necessary consequences of a pure hypothesis" (5.171).

The two periods by no means exhibit two distinct theories of abduction. The second position certainly represents Peirce's mature judgment on the matter, but it is the logical consequence of the earlier theory and can only be understood clearly in the light of the earlier theory. What follows is an attempt to trace the development of Peirce's theory of abduction.

PART I.
THE EARLY THEORY (1859-1890)

1. PEIRCE'S EARLIEST CONCEPTION OF INFERENCE (1859 - 1861)

Peirce liked to date his interest in logic from his thirteenth year when he discovered Whately's *Logic* [1] and mastered its content in a few days. Precisely when the trichotomy of reasoning entered his mind is not clear. He said in his later years that the question had been prominently in his mind since 1860 (7.97) or the early sixties (8.385). Evidently the trichotomy was a logical consequence of his discovery of the correlation between three kinds of inference and three figures of syllogism in 1865 and the discovery of the irreducibility of the three syllogistic figures in 1866.

According to Peirce, his first statement of the theory of abduction is contained in the paper "On the Natural Classification of Arguments" which he read in 1867 to the American Academy of Arts and Sciences (2.461). A fuller treatment of the theory is contained in the *Journal of Speculative Philosophy* essays of 1868. Before we turn to investigate these articles, however, it is necessary to go back to Peirce's earlier papers in order to trace the earliest development of the theory.

The earliest phase of Peirce's thought (from the earliest of his papers until 1865) was very much a Kantian phase based on Kantian logic. One of the most important principles in Peirce's theory of knowledge which he derived from Kant is the doctrine

[1] Most of Peirce's earliest papers are not published; however, a treatment of this phase of Peirce's thought is contained in Part I of Murray Murphy's book *The Development of Peirce's Philosophy,* in which he makes extensive use of the collection of unpublished Peirce manuscripts in Houghton Library at Harvard University. Murphey's book is the main source of this chapter.

that every cognition involves an inference. According to Kant there is no cognition until the manifold of sense has been reduced to unity. This reduction is accomplished by introducing a concept which is not itself a sensuous intuition. Hence cognition requires some operation upon the manifold to bring it to unity, and Peirce wrote in 1861, "An operation upon data resulting in cognition is an inference."[2]

Peirce's conception of inference is shown more clearly in his theory of perceptual judgment: "Every judgment consists in referring a predicate to a subject. The predicate is thought, and the subject is only thought-of. The elements of the predicate are experiences or representations of experience. The subject is never experiential but only assumed. Every judgment, therefore, being a reference of the experienced or known to the assumed or unknown, is an explanation of a phenomenon by a hypothesis, and is in fact an inference."[3]

Peirce regarded all mental processes as inferential. Thus "inference" includes not only deduction and induction but also *hypothesis*,[4] which is "an operation upon data resulting in cognition" or "an explanation of a phenomenon by an hypothesis." However, in his early papers Peirce did not regard these as three distinct and irreducible forms of inference. His position was that all forms of inference may be reduced to Barbara. He wrote in 1860, "It is clear that we can draw no other inference from a thing's being in a class other than what is directly expressed by Barbara namely *that whatever is true of an entire class is true of every member of the class*; hence all other syllogisms may be reduced to Barbara."[5]

From the two doctrines, viz. that all cognition is inferential, and that all inference is reducible to Barbara, it follows that every cognition is derived from a major premise and a minor premise, a major premise and a minor premise, and so on until

[2] *Ibid.*, p. 21.

[3] *Ibid.*

[4] Since Peirce only used "hypothesis" or "hypothetical inference" to designate the third kind of inference in his early period, his own usage of terms in different periods shall be preserved in this thesis so that we may follow the development of his theory more easily.

[5] Murphey, *op. cit.*, pp. 21f.

we come to some original premises ("primal truths") which are not themselves cognitions, or we have an infinite regress. What are the origins of these premises? Some may come from experience; but, since Barbara requires a universal premise and experience without cognition cannot be universal, the original major premise cannot be derived from experience alone. Thus Peirce concluded that it is only minor premises that can come from experience, major premises exist and have their truth in the mind.

The crucial problem concerning the truth of these innate premises at once arises. Since Peirce had not yet recognized induction as an autonomous form of reasoning, the possibility of the inductive confirmation of the premises was ruled out. He found himself confronted by the conclusion that the "primal truths" cannot be demonstrated, but must be accepted on faith. Peirce was understandably not happy with such a solution, however. It was rejected after 1865 as a result of his discovery of the irreducibility of the three forms of inference.

2. THREE KINDS OF INFERENCE AND THREE FIGURES OF SYLLOGISM (1862 - 1867)

Kant derived inferential forms from different propositional forms – hence there are distinct hypothetical, disjunctive, and categorial syllogism. The study of the Scholastics, however, convinced Peirce that the classification of inferential forms ought to be based upon genuine distinctions among the rules of inference or syllogistic figures themselves. The rejection of Kant's classification of propositional forms led to the reduction of the dialectical syllogism to hypotheticals.

Probably in mid-1865, Peirce came to the conclusion that every syllogism can be put into an hypothetical form; thus the first figure:

Y is X		If Y then X
Z is Y	becomes	but Y (under Z)
Z is X		Therefore X (under Z)

[6] Vide *ibid.*, p. 59.

and the second figure:

Y is X		If Y then X
Z is X	becomes	but X (under Z)
Z is Y		Therefore Y (under Z)

Similarly the third and fourth figures can be translated into hypotheticals. By "Y (under Z)" Peirce meant "Y is true if Z is." Since Peirce considered the subject as representing the cause of the predicate, "If X then Y" thus means "X is the cause of Y." This causal theory of the proposition, combined with the conversion of categoricals into hypotheticals, resulted in the following interpretation of syllogism: "Arguments in the first figure are *a priori*... Arguments in the second figure are *a posteriori*... The two kinds of arguments are alike as to their major premise, they differ in their minor premise which in the one case is the statement of a cause and in the other an effect..."[7] The third and fourth figures are hypothetical in their conclusions and the third is only probable.

It seems the correlation between the three kinds of inference and the three figures would be a logical consequence of the above analysis. However, Peirce has left an account of the discovery in which he states that it was occasioned by his endeavour to define "induction" syllogistically after reading Boole's *Laws of Thought.* He found that induction "would be defined as the *inference* of the major premise of a syllogism from its minor premise and conclusion. Now this was exactly what Aristotle said it was the 23rd Chapter of the second book of the *Prior Analytics*... With this hint as to the nature of induction, I at once remarked that if this be so there ought to be a form of inference which infers the minor premise from the major and the conclusion... I looked on further and found that... Aristotle opens the 25th (Chapter) with a description of the inference of the minor premise from the major and the conclusion..."[8]

[7] From a fragment "On the Distinction between *A priori* and *A posteriori*" in *ibid.*, p. 59.

[8] Lowell Lectures, 1903, Vol. I, no. 8, pp. 12-16. Vide Murphey, *op. cit.*, p. 60.

It should be noted that Peirce assumed all propositions to be fundamentally of the subject-predicate type. This assumption together with his causal theory of the proposition and the correlation of inferential and syllogistic forms led him to conclude:

> Reasoning *a priori* [deduction] is inferring an effect from its cause. Reasoning *a posteriori* [hypothesis] is inferring a cause from its effect. There is clearly a third way of reasoning, which consists in inferring a mutual connection between the several effects of one cause; this is *induction* (or *analogy*)...All arguments in the first figure are really *a priori,* for in the major Y is made the subject and hence the cause and in the minor it is predicated, whence the effect is predicated. So...the train of thought is from cause to effect. All arguments in the second figure are really *a posteriori*...All arguments in the third figure are really inductive...[9]

In his first period Peirce held that all forms of inference may be reduced to Barbara. The above interpretation of the three figures, however, appears to show that there are genuine differences between the principles of the three figures. Early in 1866, therefore, Peirce made a detailed analysis of the relation among the syllogistic figures, and he succeeded in proving that each figure involves an independent principle of inference. For although any syllogism of the second or third figure can be reduced to one of the first, the argument by which this reduction is made must be in the figure from which it is being reduced. "Hence, it is proved that every figure involves the principle of the first figure, but the second and the third figures contain other principles, besides" (2.807). Consequently, Peirce came to recognize the three kinds of inference as autonomous forms proceeding from three different principles.

The result of this discovery is contained in the paper "On the Natural Classification of Arguments" which was presented at the April 9, 1867, meeting of the American Academy of Arts and Sciences, (2.461). In this paper Peirce correlates the three forms of inference with the three figures of syllogism. In the

[9] *Ibid.*, p. 61.

syllogism:

$$\begin{array}{l} \text{Any M is P,} \\ \quad \Sigma\text{'S' is M;} \\ \therefore\ \Sigma\text{'S' is P;} \end{array}$$

where Σ' S' denotes the sum of all the classes which come under M, if the minor premise and conclusion are known to be true, the major premise is, by enumeration, true. This is called formal induction. The form of this inference is:

$$\begin{array}{l} \Sigma\text{'S' is P,} \\ \Sigma\text{'S' is M;} \\ \therefore \text{M is P.} \end{array}$$

In a similar way, from the syllogism:

$$\begin{array}{l} \text{Any M is } \Pi\text{'P',} \\ \text{Any S is M;} \\ \therefore \text{Any S is } \Pi\text{'P';} \end{array}$$

where Π' P' denotes the conjunction of all the characters of M, if the major premise and conclusion are true, the minor premise is true by definition. This is called formal hypothesis. The form of this inference is:

$$\begin{array}{l} \text{Any M is } \Pi\text{'P',} \\ \text{Any S is } \Pi\text{'P';} \\ \therefore \text{Any S is M.} \end{array}$$

Of all possible propositions in either of the forms:

$$\Sigma\text{' S' is M, and M is } \Pi\text{' P',}$$

one half are true. In a false proposition, some finite ratio of the S's of P's are not true subjects or predicates. Hence, of all the propositions of either of these forms which are partly true, some finite ratio, more than one half, are wholly true. The formulae of probable inference can be obtained by substituting S' for Σ' S' and P' for Π' P' in the above formulae for formal induction or hypothesis. Hence the formulae are:

Induction : S', S", S''' etc. are taken at random as M's,
S', S", S''' etc. are P;
∴ Any M is probably P.

Hypothesis : Any M is, for instance, P', P", P''' etc.
S is P', P", P''' etc;
∴ S is probably M.

Thus induction is defined as "argument which assumes that a whole collection, from which a number of instances have been taken at random, has all the common characters of those instances; hypothesis, as an argument which assumes that a term which necessarily involves a certain number of characters, which have been lighted upon as they occurred, and have not been picked out, may be predicated of any object has all these characters" (2.515).

3. AMPLIATIVE INFERENCE AND COGNITION (1868 - 1870)

The early theory of abduction received a fuller treatment in the *Journal of Speculative Philosophy* essays of 1868 (5.213-5.357). In these essays Peirce formulates his theory of cognition in accordance with his theory of signs. The single main feature in this theory is the denial of all intuitive cognition. The distinction between intuition and conception is formulated in terms of their representative character; thus, 'intuition' signifies "a cognition not determined by a previous cognition of the same object, and therefore so determined by something out of consciousness" (5.213), while a concept is always determined by a previous cognition. Peirce then argues at some length that every cognition is determined by a previous cognition of the same object, and concludes that there is no such thing as intuition.

The denial of intuition is Peirce's boldest stroke against Cartesianism and British Empiricism, for both assume the dyadic (object-subject) theory of cognition and require the existence of intuition (of "innate ideas" in the former and "sense data" in the latter) as an axiom. In denying this, Peirce had the methods of science in mind as a model. Modern science accepts no proposition as self-evident but rests on the consensus of the community of investigators as to what premises one may adopt for the sake of inquiry. Peirce thinks the existence of intuition involves the existence of the transcendental object which is absolutely inexplicable. The only way to know that something is inexplicable is by reasoning from signs. "But the only justification of an inference from signs is that the conclusion explains the fact. To suppose the fact absolutely inexplicable, is not to explain it, and hence this supposition is never allowable" (5.265). Thus Peirce considers it a postulate of logic that everything is explicable.

The problem now is to find a way to account for the so-called first cognition. It should be recalled that when Peirce asserted the existence of innate ideas in 1861, he did so on the basis of the premises that all mental processes are inferential, and that all inference is reducible to Barbara. The first of these premises he continued to hold: illation is simply a form of sign translation. But he rejected the second premise when he discovered the irreducibility of the three forms of inference. With the inclusion of ampliative inference among the forms of sign translation, Peirce attempts to account for all the forms of statement which we have without postulating innate principles. Two separate types of processes are required here: those of sign introduction and those of sign translation. But the denial of intuition means that every cognition is determined by a previous one, and the introduction of a new term may therefore be regarded as the result of hypothetical inference. Thus, for Peirce, color is a concept which is applied to the manifold of impressions as an explaining hypothesis; it is not therefore an impression itself. (For Peirce the term "impression" is restricted to the instantaneous neurological stimuli.) From the denial of intuition in general the denial of particular intuition of one's self and of the subjective elements of cognition follows. Peirce argues that the conception of self may be regarded as an hypothesis to explain ignorance and error, and no faculty of introspection need be postulated (5.225f). Hence, hypothesis seems to account for perceptual judgments and for the introduction of minor premises in general. The introduction of a new universal statement to serve as a major premise may be regarded as the result of induction, and deduction may then account for the conclusions derived.

In this period Peirce regards both induction and hypothesis as species of "reduction of a manifold to unity" (5.276). Induction is defined as "an argument which assumes that the whole collection of which a number of instances have been taken at random, has all the common characters of those instances" (2.515). This might be called statistical argument. In the long run, this kind of inference generally affords fairly correct conclusions from the premises. For example, if we have a bag of beans partly black and partly white, we may approximate the relative proportion of the two colors in the whole bag by determining the relative proportion in several handfuls taken randomly. The key to induction is that "by taking the conclusion so reached as major premise of a syllogism, and the proposition stating that such and such objects are taken from the class in question as the minor premise, the other premise of the induction will follow from them deductively" (5.274). Accordingly, induction is the inference of the major premise of a syllogism from its minor premise and conclusion. Suppose we examine certain English books at random and find that they consist of about 11.25% e's, 8.5% a's, etc.; we conclude that all English books have about 11.25% e's, 8.5% a's, etc. This is a piece of induction.

Hypothesis is defined as an argument "which proceeds upon the assumption that a character which is known necessarily to involve a certain number of others, may be probably predicated of any object which has all the characters which this character is known to involve" (5.276). Hypothesis may be regarded as the inference of the minor premise of a syllogism from the other two propositions. Suppose, upon examining a piece of writing in cipher we find that it contains less than twenty-six characters, one of which occurs about 11% of the time, another 8.5%, etc.; when we substitute for these the letters e, a, etc. respectively, we are able to make out the meaning of the cipher. We infer with great probability that the cipher makes sense in English. The function of hypothesis is "to substitute for a great series of predicates forming no unity in themselves, a single one which involves them all" (5.276). It is, therefore, like induction, a reduction of a manifold to unity.

Later in his life (1910) Peirce referred to his formulations of hypothesis and induction here and wrote "I still consider that it had a sound basis. Only in almost everything I printed before the beginning of this century I more or less mixed up Hypothesis and Induction..." (8.227). Since in his later period "hypothetical reasoning" means "the process of postulation" and "induction" means the confirmation of a postulate, it is easy to see why Peirce considered his formulation here as "mixed up." According to the present usage, both "hypothetical" and "inductive" reasoning are concerned with the process by which we create postulates. The difference between them has to do with the way in which this postulate is obtained and the kind of postulate which is obtained.

4. INDUCTION AND HYPOTHESIS (1871 - 1878)

Peirce's philosophy in the 1860's is based very largely upon the notions of classical logic, and particularly upon the subject-predicate theory of the proposition. But the discovery of the logic of relations in the late 1860's introduces propositions which are not reducible to subject-predicate form. In 1870 Peirce published his first paper on the logic of relations and analyzed the syllogisms as one form of logical relation, rather than as the fundamental formula of all argument (3.66). However, in "Deduction, Induction, Hypothesis" (2.619-644) the forms of induction and hypothesis are set up in a way similar to that of 1868. Induction is the inference of the *rule* (major premise) from the *case* (minor premise) and *result* (conclusion), while hypothesis is the inference of a *case* from a *rule* and a *result.* The following example will show the relationships more clearly:

Deduction: Rule – All the beans from this bag are white
Case – These beans are from this bag
∴ Result – These beans are white

Induction: Case – These beans are from this bag
Result – These beans are white

∴ Rule – All the beans from this bag are white
Hypothesis: Rule – All the beans from this bag are white
Result – These beans are white
∴ Case – These beans are from this bag (2.623)

"Induction is where we generalize from a number of cases of which something is true, and infer that the same thing is true of the whole class. As, where we find a certain thing to be true of a certain proportion of cases and infer that it is true of the same proportion of the whole class" (2.624). Hypothesis is where we find some surprising fact which would be explained by supposing that it was a case of a certain general rule, and thereupon adopt that supposition. The sort of inference is called "making a hypothesis" (2.623). In this kind of inference one should keep in mind that "When we adopt a certain hypothesis, it is not alone because it will explain the observed facts, but also because the contrary hypothesis would probably lead to results contrary to those observed" (2.628). Peirce seems to hint here that hypothesis *selection* is involved in this kind of inference.

Explanatory hypotheses may be of widely different kinds and Peirce alludes to at least three: (1) The kind which refers to facts unobserved when hypotheses are made, but which are capable of being observed. For example, upon entering a room I find many bags containing different kinds of beans. On a table there is a heap of white beans; I may adopt the hypothesis that the heap was taken out from a bag which contained white beans only. (2) There are hypotheses which are incapable of being observed. This is the case about historical facts. "Fossils are found, say, remains like those of fishes, but far in the interior of the country. To explain the phenomenon, we suppose the sea once washed over this land." And, "Numberless documents refer to a conqueror called Napoleon Bonaparte. Though we have not seen the man, yet we cannot explain what we have seen, namely, all these documents and monuments, without supposing that he really existed" (2.625). (3) Finally hypotheses refer to entities which in the present state of knowledge are both factually and theoretically unobservable. The kinetic theory of gases is an illustration of this kind. These are the most important kinds of hypotheses in sciences.

In order that the process of making an hypothesis should lead to a probable result, Peirce lists three rules which must be followed :

> 1. The hypothesis should be distinctly put as a question, before making the observations which are to test its truth. In other words, we must try to see what the result of predictions from the hypothesis will be.
>
> 2. The respect in regard to which the resemblances are noted must be taken at random. We must not take a particular kind of prediction for which the hypothesis is known to be good.
>
> 3. The failure as well as the success of the predictions must be honestly noted. The whole proceeding must be fair and unbiased (2.834).

The first rule comes very close to Peirce's later theory of abduction, where the three modes of reasoning are considered as three stages of inquiry. Here Peirce has not yet included methodological processes in the concept of "inference;" therefore, "prediction from the hypothesis" is not identified as deduction, and "the observations which are to test its truth" are not considered as the task of induction. The other two rules are in fact concerned with induction rather than hypothesis. Abduction proper, the process of adopting an hypothesis, is barely touched upon here. This is due to the fact that Peirce regards 'inference' as essentially an evidencing process in this period.

According to the present theory induction and hypothesis are separate forms of inference. "The essence of an induction is that it infers from one set of facts to another set of similar facts, whereas hypothesis infers from facts of one kind to facts of another" (2.642). It is impossible to infer hypothetical conclusions inductively. However, it should be noted that even in this early formulation Peirce was not prepared to separate the two forms of inference absolutely. "When we stretch an induction quite beyond the limits of our observations, the inference partakes of the nature of hypothesis" (2.640). Induction and hypothesis, therefore, may be conceived as occupying opposite ends of the continuum of ampliative inference. In the later period Peirce stretches the concept of induction to include induction of characters, and abduction will appear to be a quite different kind of inference.

5. THE METHOD OF METHODS (1879 - 1890)

It is evident from the *Popular Science Monthly* series that Peirce is concerned more and more with *inquiry* and its *methods*. Though other definitions of logic occur in his writings, the one he used in his teaching at Johns Hopkins University (1879-1884)[10] was that it is the art of devising methods of research, "the method of methods" (7.59). This conception of logic differs significantly from the traditional views and some later logicians are greatly indebted to it. Dewey remarks in his *Logic* that "The readers who are acquainted with the logical writings of Peirce will note my great indebtedness to him in the general position taken. As far as I am aware, he was the first writer on logic to make inquiry and its methods the primary and ultimate source of logical subject matter." [11]

Peirce maintains that the superiority of modern science over that of the ancients is due to a better *logic*. He made the same point in 1877 when he said "each chief step in science has been a lesson in logic" (5.363). Peirce thinks "Modern methods have created modern science; and this century... has done more to create new methods than any former equal period" (7.61). Thus Peirce considers the nineteenth century as "the age of methods" (7.62) and maintains that the producing of a method for the discovery of methods is one of the main problems of logic (3.364). This conception of logic, as Dr. Fisch points out, "is suggestive of Whitehead's dictum: 'The greatest invention of the nineteenth century was the invention of the method of invention.' "[12]

[10] Max H. Fisch & J. I. Cope, "Peirce at the Johns Hopkins University," in *Studies in the Philosophy of C. S. Peirce* (Wiener and Young, eds.), pp. 283f.

[11] John Dewey, *Logic* (New York: Henry Holt, 1938), p. 9, n. 1.

[12] Max Fisch (ed.), *Classic American Philosophers* (New York: Appleton-

One of the values Peirce saw in logic was the adapting of the methods which proved fruitful in one science to the investigations of another (7.66, 2.110). That is what the greatest progress in science has consisted in. For example: Darwin adapted in biology the methods of Malthus and the statistical method of the economists; Maxwell adapted to the theory of gases the methods of the doctrine of chances (7.66, 5.364). Now although one need not have the theory of a method in order to apply it, yet in order to adapt, with suitable modification, to his own science the method of another, an acquaintance with the principles upon which it depends will be of the greatest benefit. Peirce's work on logic may be interpreted as an attempt to provide a theory of methods for the further advancement of science. He realized that he was a pioneer in this field and remarked, "it must be confessed that we students of the science of modern methods are as yet but a voice crying in the wilderness, and saying prepare ye the way for this lord of the sciences which is to come" (7.63).

Since the scope of logic is extended so that it becomes "the method of methods," the theory of the economy of research thus becomes a part of logic. As early as 1878 Peirce had written a paper on the subject (5.601, 7.159f). In general the doctrine of economy "treats of the relations between utility and cost. That branch of it which relates to research considers the relation between the utility and the cost of diminishing the probable error of our knowledge. Its main problem is, how, with a given expenditure of money, time, and energy, to obtain the most valuable addition to our knowledge" (7.140). It is evident that the consideration of economy is very important in logic, especially in abduction, where it becomes the leading consideration in hypothesis construction and selection. However, Peirce did not recognize the intimate relations between his theory of the economy of research and his theory of abduction until after 1891. This is due to his conception of the three forms of inference as distinct types of evidencing processes only, instead of methodological processes.

The logical consequence of Peirce's conception of logic as "the

Crofts, 1951), p. 34.

method of methods" is to regard the three kinds of inference as three stages of inquiry, but this did not come about until much later. His treatment of abduction as presented in "A Theory of Probable Inference" of 1883, is fundamentally similar to that of 1878. However, the theory of probability developed in 1878 is fully manifested here and the concept of induction is stretched. With the figures of the syllogism no longer fundamental in determining types of argument, Peirce now becomes concerned with new forms of classification for probable inferences. He developed for the first time the notions of "probable deduction" and "statistical deduction." His analysis of these forms of deduction presents a complicated problem which is beyond the scope of the present study.

The forms of induction and hypothesis are set up in a way similar to his earlier formulations. The form of induction is:

> S', S", S"', etc. form a numerous set taken at random from among the M's,
> S', S", S"', etc. are found to be – the proportion ρ of them – P's;
> Hence, *probably* and *approximately* the same proportion, ρ, of the M's are P's (2.702).

When the ratio π is unity or zero, the inference is an ordinary induction, and Peirce extends the term "induction' to include all such inferences, whatever be the value of π.

Corresponding to induction we have hypothesis:

> M has, for example, the numerous marks P', P", P"', etc.,
> S has the proportion *r* of the marks P', P", P"', etc.;
> Hence, probably and approximately, S has an *r*-likeness to M (2.706).

For example, we know the ancient Mound-builders of North America present some degree of resemblance to the Pueblo Indians in all those respects in which we have been able to make comparison. We may infer, then, that in all respects there is about the same degree of resemblance between these races. In the extended sense of "induction" this kind of argument is simply an induction respecting qualities instead of respecting things. He calls this kind of induction hypothetic inference or, briefly, *hypothesis.* In practice it is "different from induction owing to the impossibility of simply counting qualities as individual things are counted" (2.706). Characters are not capable of strict enumeration, but have to be estimated.

In science, the discovery of laws is accomplished by induction; the discovery of cause is accomplished by hypothesis; and deduction is concerned with the prediction of effects. Generally speaking, "the conclusions of Hypothetic Inference cannot be arrived at inductively, because their truth is not susceptible of direct observation in single cases. Nor can the conclusions of Inductions, on account of their generality, be reached by hypothetic inference" (2.714).

At this point it would be helpful to quote Peirce's later reflections on his theory of abduction of 1883. He wrote in 1902:

> In what I there said about "Hypothetic Inference" I was an explorer upon untrodden ground... According to my own principles, the reasoning with which I was there dealing could not be the reasoning by which we are led to adopt a hypothesis,... But I was too much taken up in considering syllogistic forms and the doctrine of logical extension and comprehension, both of which I made more fundamental than they really are. As long as I held that opinion, my conceptions of Abduction necessarily confused two different kinds of reasoning. When, after repeated attempts. I finally succeeded in clearing the matter up, the fact shone out that probability proper had nothing to do with validity of Abduction... (2.102).

What he succeeded in clearing up is the notion of abduction as the reasoning that leads to the adoption of an hypothesis on probation and of induction as the testing of the hypothesis. Thus the validity of abduction is an entirely different question from that of induction. The distinctions between induction and hypothesis made in the early period, however, are kept, in a modified form, as that between Quantitative and Qualitative induction in the later period (2.755, 6.526). How Peirce cleared up his confusion of the early period will become evident in the next section.

PART II.

THE LATER THEORY (1891-1914)

1. THE TRANSITIONAL PERIOD (1891 - 1898)

Peirce's theory of abduction underwent a fundamental change during the decade between 1890 and 1900. Although the notion of abduction as the process of entertaining a hypothesis became quite explicit in the early 1890's, the three kinds of reasoning were not regarded as the three stages of inquiry until a decade later. This change does not constitute a sudden abandonment of one view in favor of another entirely different, for the change was gradual and the roots of the later view go further back. As indicated previously, Peirce's conception of logic as the method of methods definitely paved the way for the development of the later view. In fact, the later theory should be regarded as the outgrowth of the early view.

The treatment of the three forms of inference in the *Monist* series (1891-93) is essentially a restatement of his earlier views. "By hypothetic inference, I mean, as I have explained in other writings, an induction from qualities" (6.145). For example, I know that the kind of man called "mugwump" has certain characteristics and holds certain opinions. Now, if I meet a man who has some of those characteristics and holds certain peculiar opinions, I am naturally led to suppose that he is a "mugwump." This is an hypothetic inference. The mind acts in a similar way. "By the hypothetic process, a number of reactions called for by one occasion get united in a general idea which is called out by the same occasion" (6.146). Induction is equated to the process of habit formation, while deduction is the process whereby the rule or habit is actualized in action. This is the way a decapitated frog *reasons* when you pinch his hind legs. The habit serves as a major premise. The pinching is his minor premise, and the conclusion is the act of jumping away (6.144, 2.711, 6.286). Peirce's concern here (in "The Law of Mind") is merely to show that the three forms of inference have analogues in psychological phenomena. The treatment here seems little more than a restatement of the point already made in 1878 (2.643). Induction, hypothesis and analogy, "as far as their ampliative character goes, that is, as far as they conclude something not implied in the premisses, depend upon one principle and involve the same procedure. All are essentially inferences from sampling" (6.40). This statement clearly belongs to his early view, for according to the later view only abductions involve additions to the facts observed. Induction never can originate any idea but simply confirms an hypothesis.

In Pierce's other writings of the early 1890's, however, characteristics of the later view of abduction are already explicit. Taking issue with the Positivists in his 1893 revision of the earlier article "On the Natural Classification of Arguments" Pierce reiterates his contention that hypothetic inference is a legitimate and independent form of inference. An hypothesis "is adopted for some reason, good or bad, and that reason is being regarded as lending the hypothesis some plausibility" (2.511 n.1). The *Monist* series can be interpreted as aiming to offer guidance in hypothesis construction and selection. His cosmological doctrines are presented as "hypotheses" (6.606) to account for the present state of the universe (as seen in the 1890's). From Peirce's arguments *for* his own theories and *against* some of the alternative theories, two very important requirements for adopting an hypothesis emerge. First, it must explain observed facts. That is to say, the facts must follow from the hypothesis deductively (6. 606). Second, it must lead to conclusions which are capable of verification (6.62). Although Pierce defined the *form* of hypothetic inference according to his early view in the *Monist* articles, in the actual construction of his cosmology he followed his later view of hypothesis. In fact, Peirce seems to be attempting to provide a scientific metaphysics, "which might serve as guides in the choice of an hypothesis" (8.109).

In a manuscript of notes for a projected, but never completed *History of Science* written probably in the early 1890's[1] Peirce

[1] It is dated in the *Collected Papers* as c. 1896, but a more accurate date is provided by Wiener in *Studies in the Philosophy*..., p. 344 n. 5. He thinks

adopts a new term, "Retroduction," to designate what he used to call hypothesis, and mentioned that this is the same as Aristotle's "abduction" (1.65).[2] "Retroduction is the provisional adoption of an hypothesis, because every possible consequence of it is capable of experimental verification, so that the persevering application of the same method may be expected to reveal its disagreement with facts, if it does so disagree" (1.68). The conception of abduction is obviously stretched to include the methodological process as well as evidencing process. He begins to consider the reasons for adopting an hypothesis. "A hypothesis is something which looks as if it might be true and were true, and which is capable of verification or refutation by comparison with facts. The best hypothesis, in the sense of most recommending itself to the inquirer, is the one which can be most readily refuted if it is false" (1.120). "The effort should, therefore, be to make each hypothesis, . . . , as near an even bet as possible" (1.121). This deals with the reasons for constructing an hypothesis. "The purpose of a theory may be said to be to embrace the manifold of observed facts in one statement, and other things being equal, that theory best fulfills its function which brings the most facts under a single formula" (7.410). Later on Peirce introduced other guiding principles for entertaining a hypothesis.

His conception of deduction and induction remained unchanged in the early 1890's. However, by 1898, deduction is clearly regarded as the process of tracing out the necessary and probable consequences of an hypothesis. Peirce wrote "Reasoning is of three kinds. The first is necessary, but it only professes to give us information concerning the matter of our own hypotheses... The second depends upon probabilities... The third kind of reasoning tries what *il lume naturale*... can do. It is really an ap-

it should be dated 1891-1892.

[2] In the later period Peirce commonly used the term *abduction* and sometimes retroduction or presumption. He seems to prefer *abduction* as the best designation. In support of his use of the term he refers to "the doubtful theory, . . . that the meaning of the 25th chapter of the 2nd book of the *Prior Analytics* has been completely diverted from Aristotle's meaning by a single wrong word having been inserted by Appellicon where the original was illegible" (8.209). In 1901 he wrote a detailed investigation of this "doubtful theory" in "The Logic of Drawing History from Ancient Documents" (7.249). See also 2.776, 2.37n, 5.144.

peal to instinct" (1.630). Induction is not yet treated as the process of testing an hypothesis. The basic idea was virtually expressed when Peirce said, reflecting the "views of Whewell" in 1893, that "progress in science depends upon the observation of the right facts by minds *furnished with appropriate ideas*" (6.604).

It is obvious that Peirce is on his way to regarding the three modes of inference as the three stages in scientific inquiry. An interesting point to be noted is the fact that up until the end of the 19th century, Peirce always listed the three modes of inference according to degrees of certainty, namely: deduction, induction, and hypothetic inference. After he came to regard them as the three stages in an inquiry the list became: abduction, deduction and induction.

Two central problems in understanding the nature of abduction emerge from the above discussion. First, Peirce's contention that abduction is really an appeal to instinct seems to be inconsistent with the conception of abduction as a form of inference. Second, Peirce fails to make clear the distinction between hypothesis construction and hypothesis selection and their relationship to abduction. We shall deal with these problems when they come up again in Sections 3 and 5.

2. THREE STAGES OF INQUIRY

The first full statement of Peirce's later theory of abduction is contained in his 1901 manuscript "On the Logic of Drawing History from Ancient Documents." When surprising facts emerge, an explanation is required. "The explanation must be such a proposition as would lead to the prediction of the observed facts, either as necessary consequences or at least as very probable under the circumstances. A hypothesis then, has to be adopted which is likely in itself and renders the facts likely. This step of adopting a hypothesis as being suggested by the facts, is what I call *abduction*" (7.202). An hypothesis adopted in this way could only be adopted on probation and must be tested. Peirce calls abduction the "First Stage of Inquiry (6.469). "The first thing that will be done, as soon as a hypothesis has been adopted, will be to trace out its necessary and probable experimental consequences. This step is *deduction*" (7.203).

The next step is to test the hypothesis by making experiments and comparing the predictions drawn from the hypothesis with the actual results of the experiment. When we find that prediction after prediction is verified by experiment we begin to accord to the hypothesis a standing among scientific results. "This sort of inference it is, from experiments testing predictions based on a hypothesis, that is alone properly entitled to be called *induction*" (7.206).

The three kinds of inference now become three stages in a scientific inquiry. They are intimately connected as a *method.* Peirce's view on the relationship between the three modes of inference remains essentially the same from this date. He confines his attention mostly to scientific reasoning and "inference" is mainly treated as methodological process. In order to show more clearly the differences between the present theory of abduction and the early theory, we shall now consider Peirce's threefold division of induction.[3]

The first and the weakest kind he calls "crude induction."[4] This is most clearly exemplified in the practice of generalizing about the trend of future events on the basis of past experience. The sun has risen every day, therefore, we presume it will continue to do so (8.237). It is practically the same as Bacon's "induction by simple enumeration." The obvious weakness of crude induction is that "if its conclusion be understood as indefinite, it will be of little use, while if it be taken definitely, it is liable at any moment to be utterly shattered by a single experience" (2.757). Hence its role in science can never be other than a minor one.

From the weakest kind of induction Peirce turns to the strongest which he terms "Quantitative Induction."[5] It investigates the interrogative suggestion of Abduction, "What is the 'real proba-

[3] Peirce repeatedly attempted to make his classification clear (7.208ff., 2.755ff.). The brief account presented here represents his more mature view on this matter.

[4] Peirce sometimes calls it "Rudimentary Induction" (7.111) or "pooh-pooh argument" (2.269).

[5] Peirce sometimes calls it "Statistical Induction" (7.120).

bility' that a member of a given experiential class will have a certain character?" The subject-matter being investigated consists of denumerable units. We first draw a sample of the given class, find a numerical expression for the pre-designated character of that sample, and then extend this evaluation, under proper qualification, to the entire class, by the aid of the doctrine of chances (7.120).

The third kind of induction stands midway between the other two in respect of both the security and the scientific value of its conclusions, but is of more general utility than either of the others. Peirce calls it "Qualitative Induction."[6] "It consists of those inductions which are neither founded upon experience in one mass, as Crude Induction is, nor upon a collection of numerable instances of equal evidential values, but upon a stream of experience in which the relative evidential values of different parts of it have to be estimated according to our sense of the impressions they make upon us" (2.759). This kind of induction tests an hypothesis by sampling the possible predictions that can be based upon it. Predictions are not units and their significance can only be estimated. Consequently, we cannot say that a collection of predictions constitutes a strictly random sample, and therefore probability is not involved in this kind of induction.

The foregoing types of induction have one important feature in common. They are "mere processes for testing hypotheses already in hand" (7.217). We are now in a better position to understand the later theory of abduction and its relation to the early theory. According to the early view both abduction and induction are "synthetic" in the sense that something not implied in the premises is contained in the conclusion. The difference between them lies in the results of the inferences. Induction is reasoning from particulars to a general law; abduction, reasoning from effects to cause. The former classifies while the latter explains. On Peirce's present view *any* synthetic proposition, whether it is a nonobservable entity or a generalization (so-called), in so far as it is for the first time entertained as possibly

[6] Also called "Abductory Induction" by Peirce, because it involves an element of guess-work owing to the vagueness of the characteristics being investigated (6.526).

true, it is an hypothesis arrived at by abduction. He states "Any proposition added to observed facts, tending to make them applicable in any way to other circumstances than those under which they were observed may be called a hypothesis... By a *hypothesis*, I mean, not merely a supposition about an observed object... but also any other supposed truth from which would result such facts as have been observed, as when Van 't Hoff, having remarked that the osmotic pressure of one per cent solutions of a number of chemical substances was inversely proportional to their atomic weights, thought that perhaps the same relation would be found to exist between the same properties of any other chemical substance" (6.524f.). On the early view this last example of an abduction would have been called a "generalization" which would only be the result of induction. On the present view such generalization is *suggested* by abduction and only *confirmed* by induction. In fact, Peirce now considers "laws" or "generalizations" explanatory hypotheses. He writes, "An explanation of a Phenomenon as the term is used in the so-called 'descriptive' sciences,... consists in showing that the observed phenomenon follows logically, either necessarily or probably, from Explanatory Hypothesis required by sound logic. (Since science begins in observation, followed by explanation, which in time leads to classification of phenomena, and classification ultimately results in the discovery of law applicable to further explanation)."[7]

Induction and abduction, according to the early view, are different ways of *inferring* different kinds of hypotheses. The early view of induction now corresponds to "Quantitative Induction" and what was formally "Hypothetic Inference" is now called "Qualitative Induction," both being ways of *confirming* hypotheses. Thus, it is evident that, on the present view, abduction alone can supply new ideas. It is the only kind of reasoning which is, in this sense, synthetic (2.77, 5.145, 5.171). Induction never can originate any idea whatever, for it simply confirms what was already tentatively contributed to knowledge. The relationship between abduction and induction is now very clear. "The induc-

[7] *Logic Notebook, C.S. Peirce Papers,* Houghton Library, Harvard University, p. 294 (1908).

tion adds nothing. At the very most it corrects the value of a ratio or slightly modifies a hypothesis in a way which had already been contemplated as possible. Abduction, on the other hand, is merely preparatory. It is the first step of scientific reasoning, as induction is the concluding step... They are the opposite poles of reason, the one the most ineffective, the other the most effective of arguments. The method of either is the very reverse of the other's... Abduction seeks a theory. Induction seeks for facts" (7.217-218).

3. ABDUCTION AND GUESSING INSTINCT

As stated in the Introduction, the main question in scientific discovery is, "Is the process of conceiving a new idea a reasonable affair which is subject to a logical inquiry or is it purely an *insight* of an investigator?" The usual view regards hypotheses in science as fruitful guesses or flashes of insight. Peirce holds the apparently paradoxical view that hypotheses are indeed flashes of insight and that they are inferences. In order to understand the nature of abduction this puzzling aspect of Peirce's theory must be resolved.

Peirce sometimes speaks of abduction as essentially a kind of guessing instinct. Abduction "tries what *il lume naturale* . . . can do. It is really an appeal to instinct" (1.630). "The abductive suggestion comes to us like a flash. It is an act of *insight,* although of extremely fallible insight" (5.181), and "abduction is...nothing but guessing" (7.219). As early as 1883 in "A Theory of Probable Inference" Peirce had already maintained: "Nature is a far vaster and less clearly arranged repertory of facts than a census report; and if men had not come to it with special aptitudes for guessing right, it may well be doubted whether in the ten or twenty thousand years that they may have existed their greatest mind would have attained the amount of knowledge which is actually possessed by the lowest idiot" (2.753). And he went on to declare that "All human knowledge, up to the highest flights of science, is but the development of our inborn animal instincts" (2.754, 6.604).

These remarks, taken at their face value, certainly seem to be inconsistent with the notion of abduction as a form of inference. In fact, this kind of remark permits a loose judgment as exemplified by Braithwaite's remarks in his review of the *Collected Papers*. He says, "Peirce only differs from the orthodox account in classifying the thinking of the hypothesis as itself a form of ampliative reasoning, and this difference is, I think, merely a verbal one, namely, that of whether (an act of *insight*...) (5.181) is or is not called reasoning."[8] If this criticism is correct, the process of conceiving new ideas cannot be subject to logical analysis and it would be more properly investigated by psychologists, historians, and sociologists. However this is clearly not an accurate account of Peirce's view. After briefly discussing the fundamental conceptions of logic and the trichotomy of inference in a partial synopsis of a proposed work, "Minute Logic" of 1902, Peirce turns to discuss what he calls Speculative Rhetoric or Methodeutic and says, "After the main conceptions of logic have been well settled, there can be no serious objection to relaxing the severity of our rule of excluding psychological matter, observations of how we think and the like... But while the justice of this must be admitted, it is also to be borne in mind that there is *a purely logical doctrine of how discovery must take place,...* [my emphasis]. In addition to this, there may be a psychological account of the matter, of the utmost importance and ever so extensive.. I may here and there make such use of it as I can in aid of my doctrine" (2.107). There is no doubt that abduction is concerned with the *logic* of discovery, but it should be understood that Peirce did not confine himself to logical problems alone; he was also interested in the *psychology* and *history* of discovery. It is true that Peirce did not always keep the distinction in his writings, but in order to have a clear understanding of Peirce's theory of abduction it is necessary for us to keep the distinction in mind.

Peirce had made this clear in his discussion of the validity of induction and abduction in 1883, when he said, "Others have supposed that there is a special adaption of the mind to the universe, so that we are more apt to make true theories than we

[8] R. B. Braithwaite, "Review of the *Collected Papers of C. S. Peirce*," *Mind*, XLIII (New Series, 1934), p. 510.

otherwise should be. Now, to say that a theory such as this is *necessary* to explaining the validity of induction and hypothesis [abduction] is to say that these modes of inference are not in themselves valid,..." (2.749). Peirce insists that abduction is valid in itself. But this fact alone does not explain the rapidity with which men hit upon right theories. Peirce was impressed by the fact that though man has been engaged in inquiry but a comparatively short time, he has somehow managed to acquire a number of true scientific doctrines. A special adaptation of the mind to the universe may be considered an *hypothesis* to account for this fact but it is *not* necessary for explaining the validity of abduction. Thus, Peirce maintains, "There can... be no reasonable doubt that man's mind, having been developed under the influence of the laws of nature, for that reason naturally thinks somewhat after nature's pattern. This vague explanation is but a *surmise*,..." (7.39); "...It is a primary *hypothesis* underlying all abduction that the human mind is akin to the truth in the sense that in a finite number of guesses it will light upon the correct hypothesis" (7.220); and "Retroduction goes upon the *hope* that there is sufficient affinity between the reasoner's mind and nature's to render guessing not altogether hopeless,..." (1.121).[9]

This hypothesis is quite consistent with the evolutionary speculations current in that day. It is adopted to explain the fact that men are able to hit upon true theories in a limited number of guesses. "You cannot say that it happened by chance, because the possible theories... exceed a trillion,... and therefore the chances are too overwhelmingly against the single true theory... ever having come into any man's head" (5.591). The affinity of the mind with nature is derived from his instinctive life through the process of evolution, for man's mind has been developed under the influence of the laws of nature. "It is to be expected that he should have a *natural light* or *light of nature*, or *instinctive insight,* or genius, tending to make him guess those laws aright or nearly aright" (5.604). Consequently, Peirce says, "It is somehow more than a mere figure of speech to say that nature

[9] My emphasis. See also 2.753, 5.173, 6.531, 1.81, 5.47, 2.796, 2.86, 1.121, etc.

fecundates the mind of man with ideas which, when those ideas grow up, will resemble their father, Nature" (5.591).

It should be clear by now that the accusation by Braithwaite concerning Peirce's account of abduction is not valid, for when Peirce talks about "insight" and "the affinity of mind with nature" he is only concerned with the psychological and historical aspects of discovery; not the logical. In fact, Peirce asserts, "You may produce this or that excellent psychological account of the matter. But let me tell you that all the psychology in the world will leave the logical problem just where it was" (5.172). This point will become ever clearer in the next section where we shall discuss Peirce's conception of logic as a Normative Science.

4. LOGIC AS A NORMATIVE SCIENCE[10]

There is much disagreement as to the nature of logic. Peirce not only regards logic as *normative* science (1.577) but even went so far as to make it dependent upon ethics and, in turn, aesthetics (1.611, 1.573, 5.35, etc.). He advanced this conception of logic late in life (around 1900. See 2.197, 5.111, 5.129). But the roots of his conception of logic as based on ethics go further back to 1868 when he maintained that logic is rooted in the social principle (2.655). As he based most of his philosophy on his logic, his views on this subject throw considerable light on his theory of abduction.

In order to understand the normative character of logic we must first look into the distinction between *logica utens* and *logica docens*. In ordinary life everybody has a reasoning instinct or habits of reasoning by which he forms his opinions concerning many matters of great importance. In applying such instincts, such habits, or all that goes by the name of the rule of thumb, we do *reason* in one sense (2.175). In fact, "It is really instinct that procures the bulk of our knowledge" (2.181). Peirce thinks that we not only have a reasoning instinct but we have an instinctive *theory of reasoning*, for every reasoner "has some general idea of what good reasoning is" (2.186). Such a theory of

[10] See Arthur Burks, "Peirce's Conception of Logic as a Normative Science," *Philosophical Review*, XII (1943), pp. 187-93.

reasoning, antecedent to any systematic study of the subject, constitutes our *logica utens,* the acritical and implicit logic of the common man (2.189, 2.204, 2.3, etc.).

But man does not possess a full stock of instincts to meet all occasions, "and so is forced upon the adventurous business of reasoning... When one's purpose lies in the line of novelty, invention, generalization, theory – in a word, improvement of the situation –... instinct and the rule of thumb manifestly cease to be applicable..." (2.178). Thus we proceed to study the processes of reasoning and inquire the methods by which we can most speedily advance our knowledge. The result of such study is called *logica docens,* or formulated, scientific and critical logic (2.186f, 2.204).

Thus, by our *logica utens* we are able to guess right in many instances. This ability, as suggested in the previous section, may be regarded as the result of the adaptation of the mind to the universe. But, where our instinctive reasoning power begins to lose its self-confidence, as when we are confronted with extraordinary and unusual problems, we look to the help of our *logica docens.* Peirce warns, however, that we should not expect the study of logic to supply an artificial method of doing the thinking that our regular business requires us daily to do (2.3), and suggests that "The best plan, then, on the whole, is to base our conduct as much as possible on Instinct, but when we do reason to reason with severely scientific logic" (2.178).

Now, "Reasoning, properly speaking, cannot be unconsciously performed... For reasoning is deliberate, voluntary, critical, controlled, all of which it can only be if it is done consciously" (2. 182). Peirce considers reasoning a species of conduct (1.610, 5.534) which is subject to criticism (7.187) in the sense of approval or blame. A mental operation which is similar to reasoning in every other respect except that it is performed unconsciously cannot be called "reasoning" (2.182) because "it is idle to criticize as good or bad that which cannot be controlled" (5.108). Since reasoning is a kind of voluntary and deliberate conduct, we are held responsible for its consequences. It is clear such conduct comes under the domain of ethics for ethics is the theory of self-controlled or deliberate conduct (1.191, 1.575).

Ethics and logic are both normative because "nothing can be either logically true or morally good without a purpose to be so. For a proposition, and especially the conclusion of an argument, which is only accidentally true is not logical" (1.575). Logic is a study of the means of attaining the end of thought. What is our ultimate aim? It is the business of ethics to find this out. Thus logic "must appeal to ethics for its principles" (1.191), and "good reasoning and good morals are closely allied" (1.576). Peirce points out, "in induction a habit of probity is needed for success... and... industry is essential. In the presumptive choice of hypothesis, still higher virtues are needed – a true elevation of soul. At the very lowest, a man must prefer the truth to his own interest and well-being... if he is to do much in science" (1.576). After showing the intimate relationship between logic and ethics, Peirce goes on to speculate that ethics may be ultimately based on aesthetics (2.99, 5.11). This novel doctrine is highly illuminating, however, it is beyond our present task to investigate it.

Once the conception of logic as normative is clear in our minds, we are in a better position to understand Peirce's definition of logic as "the theory of *right* reasoning, of what reasoning ought to be" (2.7). "Logic is not the science of how we *do* think; but... how we *ought* to think; nor how we ought to think in conformity with usage, but how we ought to think in order to think what is *true*" (2.52). This is analogous to ethics, which is concerned with what we *ought* to do, not what we in fact do. We are held responsible for our reasonings, just as we are morally responsible for our conduct.

This conception of logic throws considerable light on the nature of abduction. It is obvious that the process of hypothesis construction and selection is a conscious, deliberate, voluntary and controlled conduct, and thus open to criticism at every step. This, however, does not imply that we must be aware of the whole process of the mind in reasoning. "All that is necessary is that we should, in each case, compare premises and conclusion, and observe that the relation between the fact expressed in the premises involves the relation between facts implied in our confidence in the conclusion" (2.183). The act of adopting an hypothesis itself, at the instant, may seem like a flash of insight, but afterward it may be subjected to criticism. When a scientist is asked, "Why do you advance such an hypothesis?" it would not do for him to reply, "It was just one of my flashes of insight." He is expected to give *reasons* for his suggestion. In fact, whenever a scientist proposes an hypothesis to account for some facts, he always attempts to furnish *reasons,* good or bad, as to why he thinks it is the best hypothesis. It is the task of the logician to analyse these reasons and to discover an ideal method of investigation for pursuing the truth.

In view of the above analysis we may now answer the question posed at the beginning of this monologue positively. Besides history, psychology, and sociology of scientific discovery, *there is a logic of scientific discovery*. Our next task is to investigate the nature of such a logic.

5. HYPOTHESIS CONSTRUCTION AND SELECTION

What does abduction consist in? Is it the logic of *constructing* an hypothesis, or the logic of *selecting* an hypothesis from among many possible ones? Peirce himself did not always keep this distinction in mind and often treated them as the same question. In some of his writings he maintains "Abduction consists in studying facts and devising a theory to explain them" (5.145); "Abduction is the process of forming an explanatory hypothesis" (5.171); or abduction "consists in examining a mass of facts and in allowing these facts to suggest a theory" (8.209). In other writings he regards abduction as "the process of choosing a hypothesis" (7.219). To understand the nature of abduction it is necessary to investigate the relationship between hypothesis construction and selection.

As noted in the previous sections, abduction is concerned with analysing the *reasons* for proposing an hypothesis. The question now is: Is abduction concerned with the reasons for constructing an hypothesis in a certain way, or is it concerned with the reasons for preferring one hypothesis over many other possible ones? At the outset these seem to be two entirely different questions, but in practice the way one constructs an hypothesis is intimately connected with the notion of choosing the best hypothesis. The purpose of constructing an hypothesis is to *explain* some facts. But for any given set of facts there may be a countless number of possible explanatory hypotheses. "Consider the multitude of theories that might have been suggested. A physicist comes across some new phenomenon in his laboratory. How does he know but the conjunctions of the planets have something to do with it or that it is not perhaps because the dowager empress of China has at that same time a year ago chanced to pronounce some word of mystical power or some invisible jinnee may be present" (5.172); or "his daughter having on a blue dress, he having dreamed of a white horse the night before, the milkman having been late that morning, and so on?" (5.591). In one sense the proposing of an hypothesis is no problem at all. But of the trillions of hypotheses that might be made only one is true. The problem of constructing a good hypothesis is, thus, analogous to the problem of choosing a good hypothesis. The two questions, in practice, merge together.

This analysis is implicit in the following definition of abduction: "The first starting of a hypothesis and the entertaining of it, whether as a simple interrogation or with any degree of confidence, is an inferential step which I propose to call abduction. This will include a preference for any one hypothesis over others which would equally explain the facts as long as this preference is not based upon any previous knowledge bearing upon the truth of the hypotheses, nor on any testing of any of the hypotheses, after having admitted them on probation. I call all such inference by the peculiar name, *abduction*..." (6.525).

What are the logical rules to which we must conform in abduction? Peirce answers, "There would be no logic in imposing rules, and saying that they *ought* to be followed, until it is made out that the purpose of hypothesis requires them" (7.202). "Now the only way to discover the principles upon which anything ought to be constructed is to consider what is to be done with the constructed thing after it is constructed" (7.220). The first thing to be done with the hypothesis is to trace out its consequences by deduction, and then compare them with the results of experiment by induction. As soon as the first one has been refuted, we modify or discard the hypothesis and try another, hoping we shall light upon the true one, at last. In view of this prospect, it is plain that the reasons for constructing an hypothesis of a certain kind are analogous to the reasons for preferring any one hypothesis over others. The central problem of abduction, therefore, is to understand the conditions or the criterion for the best hypothesis.

Peirce names three main considerations that should determine our choice of an hypothesis (7.220). In the first place, an hypothesis must be such that it will explain the surprising facts we have before us. In the second place, it must be capable of being subjected to experimental testing. This point is closely connected with the doctrine of Pragmatism. "In the third place, quite as necessary as consideration as either of those I have mentioned, in view of the fact that the true hypothesis is only one out of innumerable possible false ones, in view, too, of the enormous expensiveness of experimentation in money, time, energy, and thought, is the consideration of economy' (7.220). These last two considerations will be discussed in more detail in the following sections. At present we shall only be concerned with the first consideration.

The whole motive of our inquiry is to rationalize certain surprising facts by the adoption of an explanatory hypothesis. "The hypothesis cannot be admitted, even as an hypothesis, unless it be supposed that it would acount for the facts or some of them. The form of inference, therefore, is this:

> The surprising fact, C, is observed;
> But if A were true, C would be a matter of course;
> Hence, there is reason to suspect that A is true.

Thus, A cannot be abductively... conjectured until its entire content is already present in the premise, 'If A were true, C would be a matter of course' " (5.189). This explanation shows how the phenomenon would be produced, come about, or result in case the hypothesis were true. It may "consist in making the observed facts natural chance results, as the kinetical theory of gases explains facts; or it may render the fact necessary" (7.220).

The negative aspect of the first consideration in abduction is that "we must not make hypotheses that will absolutely stop inquiry" (7.480). Peirce has consistently held, since 1868, that logic requires us to postulate of any given phenomenon, that it is capable of rational explanation (5.265). It is never allowable to suppose the facts absolutely inexplicable. Nothing justifies our abduction except its affording an explanation of the facts. It is no explanation at all to pronounce a fact inexplicable (1.139, 1.170, 8.168). That blocks the way of inquiry. Thus, Peirce urges us to inscribe upon every wall of the city of philosophy the slogan: "Do not block the way of inquiry" (1.135).

6. ABDUCTION AND PRAGMATISM

What should an explanatory hypothesis be to be worthy to rank as an hypothesis? Of course, as shown above, it must explain the facts. But what other conditions ought it to fulfill to be good? A very important condition is that it must be capable of experimental verification. In this respect Abduction has a close and significant connection with pragmatism. Since the former is the process of formulating hypotheses, and the latter is a doctrine whereby meaningful conceptions are distinguished from meaningless, it is not hard to see that from one point of view pragmatism is simply "the logic of abduction." "That is, pragmatism proposes a certain maxim which, if sound, must render needless any further rule as to the admissibility of hypotheses to rank as hypotheses, that is to say, as explanations of phenomena held as hopeful suggestions; and, furthermore, this is *all* that the maxim of pragmatism really pretends to do, at least so far as it is confined to logic..." (5.196).

According to the pragmatic maxim only that hypothesis is admissible which is meaningful, in the sense that it is capable of experimental verification; if two or more hypotheses yield the same practical consequences, then their logical import is identical, no matter how much their verbal expression may differ. The propositions which the maxim admits, all philosophers would agree, ought to be admitted; while, on the other hand, it cannot exclude any hypothesis which ought to be considered. For all the maxim asks is that the proposed hypothesis should have some empirically testable effects. The criticism it applies does not rule out the highest flights of imagination, "provided this imagination ultimately alights upon a possible practical effect; and thus many hypotheses may seem at first glance to be excluded by the pragmatical maxim that are not really so excluded" (5.196).

On the matter of the admissibility of hypotheses to rank as hypotheses pragmatism differs significantly from positivism. Comte and Poincaré, in common with the majority of positivists, embrace a "descriptive" theory of science, according to which the propositions of science should properly *describe* sense-impressions. For them, the function of empirical laws is to describe the world, and the laws presumable *are* verified by direct perception. An hypothesis is simply a conceptual device for stimulating and directing observation; it is not regarded as the result of an inference. For this reason, hypothetical constructions ought to be used sparingly by scientists and philosophers. Indeed, Comte goes so far as to contend that no hypothesis is admissible which is not capable of verification by *direct observation*. On such a view unperceived elements would have to be rigorously excluded from scientific knowledge (1.138, 2.511 n.1, 5.597, 7.91, 7.203).

Peirce regards the positivist canons for hypothesis as a reflection of the nominalistic tendency of Comte's time. The function of science, Peirce maintains, is essentially *explanatory*. Empirical laws are a kind of hypothesis which are adopted to *explain* facts by showing that the observed phenomenon follows logically, either necessarily or probably, from them. The only condition pragmatism imposes on an hypothesis, even if it involves assumed entities, is that it be capable of experimental verification. The dictum of Comte leads to consequences which are contradictory to the actual results of science. For instance, it would forbid an archeologist who uncovers a deposit of ancient arms and utensils to suppose that they were either made or used by any human beings, since no such beings could ever be detected by direct observation. "The same doctrine would forbid us to believe in our memory of what happened at dinnertime today," and "with memory would have to go all opinions about everything not at this moment before our senses" (5.597). In short, this whole doctrine reduces to an extreme phenomenalism, and leads ultimately to "solipsism of the present moment."

Peirce's full support of hypotheses which posit unobserved or unobservable entities is based on his examination of the actual procedure of science and his long experience as an experimental physicist. In his review of the first volume of James' *Principles of Psychology* he takes issue with the dictum in James' Preface that "all attempts to explain our phenomenally given thoughts as products of deeper lying entities... are metaphysical". Pointing out that the application of the epithet "metaphysical" is merely an indication of the author's personal distaste which signifies nothing. Peirce says, "Nor is it in the least true that physicists confine themselves to such a 'strictly positivistic point of view.' Students of heat are not deterred by the impossibility of directly observing molecules from considering and accepting the kinetical theory; students of light do not brand speculations on the luminiferous ether as metaphysical... All these are 'attempts to explain phenomenally given elements as products of deep-lying entities.' In fact, this phrase describes, as well as loose language can, the general character of scientific hypothesis" (8. 60). The important consideration is that hypotheses of this sort do have observable consequences and may be verified. *Any* statement is meaningful if it is capable of confirmation.

Peirce points out that the end of an hypothesis is, "through subjection to the test of experiment, to lead to the avoidance of all surprise and to the establishment of a habit of positive expectation that shall not be disappointed" (5.197). "When, however, we find that prediction after prediction, notwithstanding a preference for putting the most unlikely ones to the test, is verified by experiment, whether without modification or with a merely quantitative modification, we begin to accord to the hypothesis a standing among scientific results" (7.206). For, according to Pragmatism, "the entire meaning of an hypothesis lies in its conditional experiential predictions: if all its predictions are true, the hypothesis is wholly true" (7.203).

Perhaps the contention that pragmatism is the "logic of abduction" will now be clearer. After all, "Abduction must cover all the operations by which theories and conceptions are engendered" (5.590). However, we should not understand Peirce to be meaning that pragmatism is the *whole* of the logic of abduction. In fact pragmatism is only concerned with the "admissibility of hypotheses to rank as hypotheses" (5.196). Any hypothesis is admissible, according to pragmatism, if it is capable of experimental verification. But which one out of a number of admissible hypotheses should be tested first involves the question of economy which is not a part of pragmatism. Thus, pragmatism should be regarded as *an essential element* of abduction instead of the *whole* of the logic of abduction.

7. ECONOMY OF RESEARCH

We have seen, from the above two sections, that an hypothesis, to be admissible, must explain the facts at hand and must be capable of experiential verification. But the number of possible hypotheses satisfying both conditions, and concerning the truth or falsity of which we really know nothing, may be very great. The problem now is to decide which one should be subjected to verification first. One might ask, "What need of reasoning [is] there? Is he not free to examine what theories he likes? The answer is that it is a question of economy. If he examines all the foolish theories he might imagine, he never will... light upon the true one" (2.776). For the testing of hypothesis is usually more or less costly in money, time, thought, and energy (5.600, 7.200). "Not infrequently the whole life's labor of a number of able men is required to disprove a single hypothesis and get rid of it" (6.530). Therefore, "the whole question of what one out of a number of possible hypotheses ought to be entertained becomes purely a question of economy" (6.528).

The theory of the economy of research is a part of logic, for Peirce has so extended the meaning of "logic" that it becomes "the method of methods" (See Part I, Section 5). Abduction, as the first stage of inquiry, is concerned with the *reasons* for adopting a hypothesis on probation. It is easy to see that one of the leading *reasons* is the consideration of Economy. There are certain conditions which ought to guide our choice of a hypothesis. They have emerged from, and have been validated by, the actual history of successful scientific inquiry. It is the logician's job to state them as clearly as possible.

Peirce names three kinds of economic factors that should be considered in abduction: cost of verifying the hypothesis, the intrinsic value of the hypothesis proposed, and its effect upon other projects (7.220). Under the head of cost, if a hypothesis can be put to the test with very little expense (in terms of money, time, energy and thought), or if we happen to be well provided with means for testing it, that hypothesis ought to be taken up for early examination, even if its apparent likelihood is very little (6.533, 7.220, 5.598). For even if it be barely admissible for other reasons, still it is an immense advantage to get rid of that hypothesis "so as to go toward leaving the field free for the main struggle" (1.120).

Under the head of the intrinsic value of a hypothesis, we must include those considerations which tend toward an expectation that the hypothesis may be true. The most important point, discussed in a number of places by Peirce, is the question of "simplicity." He accepts Ockham's razor[11]: *Entities should not be multiplied beyond necessity,* as "a sound maxim of scientific procedure" (5.60) and "a sound economic principle which ought to guide the scientific metaphysician" (6.535). He explains the principle as follows: "Before you try a complicated hypothesis, you should make quite sure that no simplification of it will explain the facts equally well" (5.60); or "Try the theory of fewest elements first; and only complicate it as such complication proves indispensable for the ascertainment of truth" (4.35).[12] There are two questions involved in Peirce's acceptance of the razor: (a) What does he mean by "simpler hypothesis"? (b) In what sense is the simple hypothesis preferable to the complicated one?

(a) In his early writings, Peirce took the "simpler hypothesis" to mean the "logically" simpler, "the one that adds least to what

[11] Cf., 4.1, 4.35, 7.410, 6.477, 7.92, etc.

[12] Other statements: "More elements must not be introduced into an hypothesis until it is absolutely proved that fewer are not sufficient" (6.535); "An hypothesis ought not to introduce complications not requisite to explain the facts" (4.1). Cf. 5.26, 6.24, 7.410.

has been observed" (6.479). Later reflection led him to say, "It was not until long experience forced me to realize that... it is the simpler Hypothesis in the sense of the more facile and natural, the one that instinct suggests, that must be preferred... I do not mean that logical simplicity is a consideration of no value at all, but only that its value is badly secondary to that of simplicity in the other sense" (6.477). By saying that the simpler hypothesis is the one containing the "fewest elements" Peirce does not mean the one asserting something about the fewest unobservable entities. A simple hypothesis is rather one "composed of a few conceptions natural to our minds" (6.10). So that in the first place, simple hypotheses contain fewer concepts; and secondly, of two hypotheses with the same number of concepts, the simpler is that which is most "instinctive," "natural" or "familiar" (2.740) to our minds.

(b) Peirce's acceptance of the principle of Ockham's razor does not imply his acceptance of the view that a simpler hypothesis is in every case truer. In fact, he believes that the evidence is often to the contrary in many fields of science (7.93). Nevertheless he maintains that the more familiar hypothesis should be "tried first" (cf. 4.1, 5.598, 6.532). The reason appears to be twofold: first, the simpler hypotheses can be most readily refuted if they are false (1.120), for "the simplest hypotheses are those of which the consequences are most readily deduced and compared with observation; so that, if they are wrong, they can be eliminated at less expense than any other" (6.532). That is to say, a simpler hypothesis is more verifiable because it would predict more and could be put to the test more thoroughly (5.598). Thus, good scientific economy prescribes that simple hypotheses shall be thoroughly tested before resorting to complicated ones. Secondly, Peirce believes that science could not hitherto have been so successful unless there is a kind of affinity between man and nature (6.476). He asserts that "it is a primary hypothesis underlying all abduction that the human mind is akin to the truth..." (7.220). While the simpler hypothesis may not be the true one, it is the best to start with for the reason that it at least furnishes the main direction so that we might light upon the correct hypothesis in a finite number of guesses.

There is another important consideration under the head of value, namely, the *likelihood* of an hypothesis. "If we know any positive facts which render a given hypothesis objectively probable, they recommend it for inductive testing" (7.220). It may in the long run promote economy to give such a hypothesis an early trial (6.534). "When this is not the case, but the hypothesis seems to us likely, or unlikely, this likelihood is an indication that the hypothesis accords or discords with our preconceived ideas (cf. 6.534, 5.599, 1.120); and since those ideas are presumably based upon some experience, it follows that, other things being equal, there will be, in the long run, some economy in giving the hypothesis a place in the order of precedence in accordance with this indication" (7.220). However, Peirce warns us, "experience shows that likelihoods are treacherous guides" (*ibid.*); "because likelihoods are mostly merely subjective, and have so little real value" (5.599). We must proceed with extreme caution and hold ourselves ready to throw a hypothesis overboard at a moment's notice from experience.

The third category of factors of economy is especially important, "because very rarely can we positively expect a given hypothesis to prove entirely satisfactory; and we must always consider what will happen when the hypothesis proposed breaks down" (7.220). The first consideration is the quality of *caution.* A very good example of abduction, which is most instructive in respect to caution, is the game of twenty questions. In this game, one party thinks of some individual object which is well known to all participants. The other party is entitled to ask twenty question which can be answered by "yes" or "no," and then guess what was thought of. The most economical procedure is to frame the questions in such a way that each question will exactly bisect the possibilities, so that "yes" and "no" are equally probable. If the questioning is skillful, the object will invariably be guessed; but if questioners guess blindly, they almost never hit upon the right answer. Thus, twenty skillful hypotheses will ascertain what a million stupid ones might fail to do (7.220, 6.529). "The secret of the business lies in the caution which breaks a hypothesis up into its smallest logical components, and only risks one of them at a time" (7.220).

Correlative to the quality of caution is that of breadth. When the hypothesis is broken into its elementary parts we should inquire how far the same explanation accounts for the same phenomenon when it appears in other objects. For, the purpose of a theory is "to embrace the manifold of observed facts in one statement, and other things being equal that theory best fulfills its function which brings the most facts under a single formula" (7.410). When we are in doubt as to which of two hypotheses ought to be tested first, we should press the hypotheses as far as we can in order to see which one can explain more facts. "It is, therefore, good economy, other things being equal, to make our hypotheses as broad as possible" (7.221).

The above list of conditions are necessary criteria for a good hypothesis. It should be remembered that the criteria do not insure the truth of the hypothesis for an hypothesis may satisfy all the conditions listed and still be false. In fact there can be no such list of conditions that if satisfied can assure the truth of an hypothesis. "For abduction commits us to nothing. It merely causes a hypothesis to be set down upon our docket of cases to be tried" (5.602). The truth or falsity of a hypothesis can only be determined by induction. The conditions we have discussed are simply important *guides* in abduction which are derived from practical experiences in successful scientific inquiry.

8. JUSTIFICATION OF ABDUCTION

What sort of justification can be claimed for abduction? Observe that abduction is the only logical operation which introduces any new ideas. Deduction *explicates* and proves that something *must* be; induction *evaluates* and shows that something *actually is* operative. But abduction merely suggests that something *may be* (may-be and may-be not) (5.171, 6.475, 8.238). "There is no compulsiveness nor tendency toward compulsiveness" (5.172) in this kind of operation. Since abduction merely offers suggestions there seems to be little room for the question of what justifies it (5.171, 8.238). In fact, Peirce offers no other justification besides what we have already mentioned in the process of elucidating the nature and the purpose of abduction. What follows is a summary of what we have already discussed.

There are two aspects to the problem of justification. The formal aspect is concerned with the *rationale* of abduction. The only justification for an hypothesis is that it *explains* the facts (1.89, 1.139, 1.170, 2.776, 6.606). Now to explain a fact is to show that it is a necessary or a probable result from another fact, known or supposed. Thus this part of the problem is simply a question of reducing any given abductive inference to a corresponding deduction. If the latter turns out to be valid, the correctness of the abduction is guaranteed.
The form of abduction:

> The surprising fact, C, is observed;
> But if A were true, C would be a matter of course;
> Hence, there is reason to suspect that A is true (5.189)

is valid because the corresponding deduction is valid:

> If A were true, C would be a matter of course,
> A is true;
> Hence, C is true.

Peirce's position on this had not wavered since the 1860's. He wrote in 1903, "among those opinions which I have constantly maintained is this, that while Abductive and Inductive reasoning are utterly irreducible, either to the other or to Deduction,... yet the only *rationale* of these methods is essentially Deductive" (5.146). However, it must be noted that to say "the justification for an hypothesis is that it *explains* the facts" is not saying much, for the statement is what "hypothesis" as used in scientific inquiry *means*.

The other aspect, the more important but more difficult one, deals with the possibility of abduction. The problems of justifying abduction in this respect were taken by Peirce in his early writings to be exactly the same as those of induction. He asserted, "All probable inference, whether induction or hypothesis, is inference from the parts to the whole. It is essentially the same, therefore, as statistical inference" (5.349). Peirce then proceeded to apply his theory of probability to the question of the validity of statistical inference. In 1902, however, Peirce wrote, "When, after repeated attempts, I finally succeeded in clearing the matter up, the fact shone out that probability proper had nothing to do with validity of abduction..." (2.102). Thus, on the later view, the problem of justifying abduction becomes entirely different from that of induction.

How is abduction possible? Peirce answers, "The validity of a presumptive adoption of a hypothesis being such that its consequences are capable of being tested by experimentation, and being such that the observed facts would follow from it as necessary conclusions, that hypothesis is selected according to a method which must ultimately lead to the discovery of the truth" (2. 781). And, "Its only justification is that its method is the only way in which there can be any hope of attaining a rational explanation" (2.777, cf. 5.145, 5.171, 5.603). In other words, Peirce wants to say that the validity of abduction depends upon the validity of the whole scientific method. But the consideration of the validity of the scientific method is intimately connected with Peirce's theory of reality which is beyond our present concern.

The above "justification" seems to be merely a restatement of of what abduction *is* instead of providing an independent "validity" for abduction. In fact, it is doubtful whether Peirce ever was satisfied with his justification of abduction. As late as 1910, he wrote, "as for the validity of [abduction], there seems at first to be no room at all for the question of what supports it,... But there is a decided leaning to the affirmative side and the frequency with which that turns out to be an actual fact is to me quite the most surprising of all wonders of the universe" (8.238).

Elsewhere Peirce tries to account for this "wonder" of the remarkable success which abduction has achieved in leading to true theories about nature. As shown above (Part II, Section 3), Peirce contends that the reasonable supposition is that man has come to the investigation of nature with a special aptitude for choosing correct theories. This facility is derived from his instinctive life through the process of evolution. Thus, the achievements of abduction are due to the fact that the human intellect is peculiarly adapted to the comprehension of the laws of nature.

In their immediacy abductions are often merely guesses; it is quite possible for us to guess incorrectly on the first few occasions. But in the long run, "before very many hypotheses shall have been tried, intelligent guessing may be expected to lead us to the one which will support all tests, leaving the vast majority of possible hypotheses unexamined" (6.530). This is the foundation upon which abductive inference rests.

Peirce's treatment of the validity of abduction is one of the most unsatisfactory features in his theory. The claim that abduction is *necessarily valid in itself* is essential to the whole theory (Part II, Section 3), but he seems unable to provide a clear-cut justification for it. The affinity of mind with nature is an hypothesis which can only be arrived at by abduction and thus must not be used to support the validity of abduction. This failure to provide an independent justification for abduction remains a difficult problem for contemporary philosophers who maintain that there is a logic of discovery.

CONCLUSION

The author has attempted to reconstruct a consistent account of Pierce's theory of abduction from his fragmentary and often apparently inconsistent writings. As we have seen, Peirce's ideas fall roughly into two periods. In the early period Peirce treated inference, and hence abduction, as an evidencing process. The three types of inference were considered independent forms of reasoning. Abduction is an inference from a body of data to an explaining hypothesis, or from effects to cause; induction, on the other hand, is an inference from a sample to a whole, or from particulars to a general law. In the later period the concept of inference is widened to include methodological process as well as evidencing process. The three kinds of reasoning became three stages of inquiry. Abduction is the process of forming or inventing an explanatory hypothesis to account for the facts. Deduction explicates hypotheses and induction consists in the process of testing them.

In Part I we focused our attention on the chronological development of the early theory, so that the later theory may be understood more clearly in the light of the earlier development. Part II contains a systematic analysis of Peirce's mature judgment on the matter. The main concern is to present his answers to the questions put forward at the beginning of this thesis; namely, "Is there a logic of scientific discovery? If so, what is the nature of such a logic?" Although Peirce sometimes speaks of abduction as essentially a kind of guessing instinct he explicitly maintains that besides the psychological account of discovery there is definitely a *logic* of discovery. This assertion becomes very clear in the light of his conception of logic as a Normative Science. Peirce conceives of logic as "the theory of *right* reasoning, of what reasoning ought to be" (2.7), not the theory of how we do reason. Since the process of proposing an hypothesis is conscious and voluntary conduct it is open to criticism at every step. There is definitely a logical problem involved. This point will become clearer after considering the following examples of abduction.

> A pretty good example of a retroduction is the case of stopping an unknown person in the streets of a large city, asking some bit of information and crediting it. What makes this a more characteristic case is that one will have selected the man to interrogate, and will have made the selection almost without being conscious of having done so. The justification of retroduction is that one *must* trust one's instincts through life or be content with a passivity that cannot content him. All positive knowledge must be reached if at all by an operation that begins with a conjecture. I do not mean that one reflects upon this at the outset, I mean that afterward when one subjects one's behavior for the day or for any other marked period to self-criticism that is how one ought to vindicate good retroduction.[13]

On the outset the act of choosing one from among many possibilities is purely instinctive; nevertheless, there is an underlying logic in the act. For, "afterward when one subjects one's behavior to self-criticism," *reasons*, good or bad, can be given to justify the particular choice.

The ability of a great mind like Einstein to *guess* right manifests itself not only in the highest flight of scientific inquiry but also in his daily life. The following anecdote told by Alexander King is a good illustration: One day Mr. King brought along with him a "particularly dopey" photographer to interview Einstein.

> While this character was reloading his camera, Mr. Einstein looked at him earnestly for a moment and said, "you are one of many children, aren't you?" "Yes, I have nine

[13] MS published in part in the *Collected Papers,* vol. 8, Book II, chapter 2. The passage quoted here occurs after 8.238.

> brothers and sisters. What makes you ask?" "I guessed it," said Mr. Einstein, "because it is always hard to survive and to get proper attention in such a turmoil of children. Members of large families... don't expect to be seen or to be heard unless they climb right into your lap . . . I suppose it is probably the ideal training ground for a news photographer."
>
> Later,...
>
> "I thought he was a mathematician," said the photographer. "I didn't know he was Sherlock Holmes too."[14]

A great scientist is always Sherlock Holmes too, for they are both masters in abductive reasoning. Holmes, who was a genius at inventing hypotheses, described the process as reasoning "backwards." As he put it, "most people, if you describe a train of thought to them, will tell you what the result would be... There are few people, however, who, if you told them a result, would be able to evolve... what the steps were which led up to that result."[15] The following example will show Holmes in action.

On their first meeting, Holmes guessed that Dr. Watson had just recently come back from Afghanistan. Dr. Watson naturally supposed that someone had told Holmes. But Holmes said:

> Nothing of the sort... From long habit the train of thoughts ran so swiftly across my mind that I arrived at the conclusion without being conscious of intermediate steps. There were such steps, however. The train of reasoning ran, "Here is a gentleman of a medical type, but the air of a military man. Clearly an army doctor, then. He has just come from the the tropics, for his face is dark,... He has undergone hardship and sickness, as his haggard face says clearly... Where in the tropics could an English army doctor have seen such hardship...? Clearly in Afghanistan." The whole train of thought did not occupy a second.[16]

[14] Alexander King, *May This House Be Safe From Tigers* (Signet, 1960), p. 235.

[15] A. Conan Doyle, *The Complete Sherlock Holmes* (New York: Doubleday & Co., 1927), Vol. I, p. 83.

A favorite pastime for Holmes, when he was in a street car, was to guess the trades of all his fellow travelers. Let us now see Peirce do a bit of Sherlock Holmes.

> Street cars are famous *ateliers* for speculative modeling. Detained there, with no business to occupy him, one sets to scrutinizing the people opposite, and to working up biographies to fit them. I see a woman of forty. Her countenance is so sinister, . . . yet with a grimace of amiability . . . An expression of servility and hypocrisy there is, too abject for a domestic; while a certain low, yet not quite vulgar, kind of education... is evinced, together with... The whole combination, although not striking at first glance, is seen upon close inspection to be a very unusual one. Here our theory declares an explanation is called for; and I should not be long in guessing that the woman was an ex-nun (7.196).

These examples show that the method of science has much in common with the method of detectives. The important thing here is that, besides the psychological problems involved, there is a logical problem. When Einstein and Holmes *guessed* the backgrounds of their friends, they felt, *afterwards,* compelled to give *reasons* for their correct hypotheses. Similarly, whenever a scientist proposes a hypothesis to account for some facts, he is expected to furnish *reasons,* good or bad, as to why he thinks it is the best hypothesis. What are and are not good reasons for adopting a hypothesis on probation is a logical matter, which may be decided on conceptual grounds. No further observations or experiments are required to settle such issues.

Although the majority of abductions appear to be mere guesses, they conform to a definite logical form and are subject to certain conditions. To be sure there can be a psychological investigation appropriate to *guessing* (or hypothesis making); nevertheless, the *validity* of such an act depends upon conditions that are not psychological, but logical. It is the task of the logician to analyse these conditions and to discover an ideal method for inquiry. Peirce's theory of abduction aims to have a clear understanding of the general pattern of reasoning which underlies the kind of inference that is involved in the above examples.

What does abduction consist in? Is it the logic of constructing a hypothesis, or the logic of selecting a hypothesis from among many possible ones. At the outset these seem to be two entirely different questions, but, as we have shown, in practice they are analogous to one another. The central problem of abduction is to analyse the conditions or the criteria for the best hypothesis. Peirce names three main considerations that should guide our choice of a hypothesis: (a) A hypothesis must explain the facts at hand. (b) It must be capable of being subjected to experimental confirmation. (c) It must be guided by economic considerations. The first two conditions, however, are not so much the statements of conditions of a *good* hypothesis; they are statements of what "hypothesis" as used in scientific method *means*.

The third consideration is a very important element in Peirce's theory of abduction. Since the number of possible hypotheses satisfying the first two conditions may be very great, we are faced with the problem of deciding which one should be tested first. Here the economy of research plays the dominant role. Peirce considers his theory of the economy of research a part of logic, for he has so extended the scope of logic that it becomes "the method of methods." Abduction, as the first stage of inquiry, is concerned with the *reasons* for proposing a hypothesis. Clearly, one of the leading *reasons* is the consideration of Economy. The significance of Peirce's inquiry in this field will become more and more clear as the years go by.

It is beyond our present task to ask whether Peirce's conception of logic is correct. If we grant the correctness of his conception of logic, then Peirce is justified in claiming that there is a logic of abduction and that the theory of the economy of research is a part of that logic. The preoccupations of contemporary philosophers of science with justification of induction, probability, hypothetico-deductive systems, and the principles of theory construction are perfectly consistent with the purpose of Peirce's investigation. However, as has been pointed out in the Introduction, most philosophers persist in ignoring Peirce's most important contribution to the logic of scientific method, namely his theory of abduction. "Nothing has so much contributed to present chaotic or erroneous ideas of the logic of science as failure to distinguish the essentially different characters of different elements of scientific reasoning; and one of the worst of these confusions, as well as one of the commonest, consists in regarding abduction and induction, taken together... as simple argument" (7.218). These words were true when Peirce uttered them at the dawn of this century and they remain true today.

Until Peirce brought abduction prominently into view it was an "untrodden ground." His treatment involves many problems and is by no means a totally satisfactory account. Near the end of his life, when he was stricken with poverty and ill health, the problem of abduction still occupied much of his attention. He wrote in 1911, "I am just now trying to get a small book written in which I positively prove just what the justification of each of the three types of reasoning really consists in,... and showing the real nature of Retroduction."[17] To our permanent loss, he did not live long enough to make it a reality. Back in 1908 Peirce wrote in his notebook, "I am one,... who has for about 58 years been trying to come to understand the nature of the different kinds of reasonings, and for the last twenty years has more and more led the life of a recluse in order to escape all distractions from that study..."[18] The fruit of that study, which we have now reviewed, must lead us to agree with his prediction, "I am as confident as I am of death that Logic will hereafter be infinitely superior to what it is as I leave it; but my labors will have done good work toward its improvement" (2.198).

[16] *Ibid.*, p. 24.
[17] *C. S. Peirce's Letters to Lady Welby* (Erwin Lieb, ed.), pp. 41f.
[18] *Logic Notebook, op. cit.*, p. 332v, Oct. 21, 1908 entry.

BIBLIOGRAPHY

Braithwaite, R. B. *Scientific Explanation.* Cambridge University Press, 1953.

—. Review of the *Collected Papers* of C. S. Peirce. *Mind,* 43 (n.s., 1934), 487-511.

Buchler, Justus. *Charles Peirce's Empiricism.* London: Kegan Paul, Trench, Truber & Co., 1939.

—. (ed.). *Philosophical Writings of Peirce.* New York: Dover, 1955.

Burks, Arthur W. Peirce's Theory of Abduction. *Philosophy of Science,* 13 (1946), 301-06.

—. Peirce's Conception of Logic as a Normative Science. *Philosophical Review,* 52 (1943), 187-193.

Cohen, Morris R. (ed.). *Chance, Love and Logic, by C. S. Peirce.* New York: Braziller, 1956.

Copi, Irving M. *Introduction to Logic.* New York: Macmillan, 1953.

Dewey, John. *Logic.* New York: Henry Holt & Co., 1938.

—. The Founder of Pragmatism: a review of the *Collected Papers* of Peirce. *New Republic,* 81 (1935), 338-339.

Doyle, A. Conan. *The Complete Sherlock Holmes.* New York: Doubleday & Co., 1927. Volume I.

Feibleman, James K. *An Introduction of Peirce's Philosophy Interpreted as a System.* New York: Harper, 1946.

Fisch, Max H. (ed.). *Classic American Philosophers.* New York: Appleton-Century-Crofts, 1951.

— and J. I. Cope. Peirce at The Johns Hopkins University. *Studies in the Philosophy of C. S. Peirce* (Wiener and Young, eds. See below).

Frankfurt, Harry G. Peirce's Notion of Abduction. *Journal of Philosophy,* 55 (1958), 593-597.

Gallie, W. B. *Peirce and Pragmatism.* Penguin, 1952.

Goudge, Thomas. *The Thought of C. S. Peirce.* University of Toronto Press, 1950.

Hanson, Norwood R. *Patterns of Discovery: An Inquiry into the Conceptual Foundations of Science.* Cambridge University Press, 1958.

—. The Logic of Discovery. *Journal of Philosophy,* 55 (1958), 1073-1089.

—. More on "The Logic of Discovery." *Journal of Philosophy,* 57 (1960), 182-188.

—. Is There a Logic of Scientific Discovery? *Current Issues in the Philosophy of Science* (Feigl and Maxwell, eds.). New York, 1959.

King, Alexander. *May This House Be Safe From Tigers.* Signet, 1960.

Kneale, William. *Probability and Induction.* Oxford University Press, 1949.

—. Induction and Transcendent Hypothesis. *Readings in the Philosophy of Science* (H. Feigl, ed.).

Lieb, Irwin C. (ed.). *C. S. Peirce's Letters to Lady Welby.* New Haven: Whittlock's, 1953.

Murphy, Murray G. *The Development of Peirce's Philosophy.* Harvard University Press, 1961.

Peirce, Charles Sanders. *Collected Papers.* Volumes 1-6 edited by Charles Hartshorne and Paul Weiss. Cambridge, Massachusetts, 1931-1935; and volumes 7-8 edited by Arthur Burks, Cambridge, Massachusetts, 1958.

Popper, Karl R. *The Logic of Scientific Discovery.* New York: Basic Books, 1959.

Reichenbach, Hans. *Experience and Prediction.* Chicago: University of Chicago Press, 1938.

Schiller, F. C. S. Scientific Discovery and Logical Proof. *Studies in the History and the Methods of the Sciences.* (Charles Singer, ed.). Oxford: Clarendon Press, 1917. Volume I.

Thompson, Manley. *The Pragmatic Philosopher of C. S. Peirce.* University of Chicago Press, 1953.

Thomas, Vincent, (ed.). *C. S. Peirce's Essays in the Philosophy of Science.* New York: The Liberal Science Press, 1957.

Wiener, Philip P. (ed.). *Values in a Universe of Chance: Selected Writings of C. S. Peirce.* New York: Doubleday Anchor, 1958.

— and Frederick H. Young, editors. *Studies in the Philosophy of C. S. Peirce.* Harvard University Press, 1952.

Wisdom, John O. *Foundations of Inference in Natural Science.* London 1952.

GPSR Compliance

The European Union's (EU) General Product Safety Regulation (GPSR) is a set of rules that requires consumer products to be safe and our obligations to ensure this.

If you have any concerns about our products, you can contact us on ProductSafety@springernature.com

In case Publisher is established outside the EU, the EU authorized representative is:

Springer Nature Customer Service Center GmbH
Europaplatz 3
69115 Heidelberg, Germany

Batch number: 10830943

Printed by Printforce, the Netherlands